T0344261

Mathematics for Physics

An Illustrated Handbook

Mathematics for Physics

An Illustrated Handbook

Adam Marsh

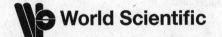

NEW JERSEY · LONDON · SINGAPORE · BEIJING · SHANGHAI · HONG KONG · TAIPEI · CHENNAI · TOKYO

Published by

World Scientific Publishing Co. Pte. Ltd.

5 Toh Tuck Link, Singapore 596224

USA office: 27 Warren Street, Suite 401-402, Hackensack, NJ 07601

UK office: 57 Shelton Street, Covent Garden, London WC2H 9HE

British Library Cataloguing-in-Publication Data
A catalogue record for this book is available from the British Library.

MATHEMATICS FOR PHYSICS
An Illustrated Handbook

ISBN 978-981-3233-91-1

For any available supplementary material, please visit
http://www.worldscientific.com/worldscibooks/10.1142/10816#t=suppl

Desk Editor: Christopher Teo

Typeset by Stallion Press
Email: enquiries@stallionpress.com

Printed in Singapore

Preface

This book is meant to complement traditional textbooks by covering the mathematics used in theoretical physics beyond that typically covered in undergraduate math and physics courses. The idea is to provide an intuitive, visual overview of these mathematical tools, with guiding end goals including but not limited to spinors and gauge theories.

What this book is, and what it is not

Because this is not a textbook, it is not designed around a traditional course or a body of specialized research results. Instead, it is a reference to help place such courses and research programs in a broader context. To make clear the somewhat unusual viewpoint taken here, we give an explicit enumeration of attributes. This book will focus on:

- Defining and visualizing concepts and relationships between them
- Stating related results and introducing related concepts for further study
- Explaining the jargon and alternative treatments found in the literature

The book will avoid:

- Historical motivations and attributions
- Proofs of theorems and derivations of results
- Tools for practical calculations

The idea is to take advantage of the reader's intuitive ability to grasp a concept through in most cases mathematically rigorous definitions, descriptive pictures, and related results. Such an approach allows us to concisely cover a large breadth of material, hopefully providing a cross-subject synthesis while at the same time serving as a useful reference.

Of course there is no doubt that a certain level of insight only comes through studying actual derivations or proofs; in particular, this is required to obtain a clear understanding of why many of the results stated here follow from the relevant definitions. Furthermore, any significant work using or extending these ideas will require tools for calculation and exercises to build facility with these tools. Thus the present book is meant as a companion to textbooks that provide these attributes.

In this book we also go out of our way to avoid messy details, special cases, and pathological exceptions. Our goal is to provide a unified presentation of core concepts so as to illuminate the assumptions and shared structures used in our models of nature. To realize this goal, the presentation strives to be consistent and intuitive. Looking up a concept here should wherever possible not only give a definition, but also place this definition in context with related concepts.

It should also be mentioned that this book in no respect attempts to be a comprehensive reference. In particular, a thin slice of the vast world of mathematics has been selected, and definitions are not necessarily the most general, the goal being coverage of structures and usage as commonly found in theoretical physics. The path chosen through theoretical physics itself reflects an end destination of theoretical particle physics, and thus omits many other branches of study.

Who this book is written for

Many people in think in terms of pictures, and then perhaps translate these pictures into equations in order to obtain concrete results. This book is written for anyone who would like to explore this viewpoint in studying (or teaching) theoretical physics.

The focus of most physics courses is on derivations of results and practice in using calculational tools. In addition, historical development is often the strongest organizing influence in presenting the material. The net result is that the student can gain facility yet lack a top-down vision of structure, instead seeing the material built up as an ad-hoc series of definitions, guesses, and leaps of faith. This is of course how the subject was developed historically, but a complementary bird's-eye view can bring together the details to form a more cohesive, and possibly more inspiring, picture of exactly what is being said about nature.

For beginning graduate and advanced undergraduate students surveying the field and choosing a specialty, the hope is that this approach will be especially helpful. In particular, with this book in hand, enough jargon may be mapped to concepts to better digest survey papers and other literature in theoretical physics.

In addition, each subject carries it's own conventions, notation, and domain of validity. An overview, presented in consistent language, can help illuminate the relationships between different subjects and make clear the common structures used across sub-fields. Sections of the book that cover material not yet learned in detail can still be useful to the student as preparation, introducing concepts and clearly demarcating the framework which future study will flesh out.

Organization of the book

The book starts with generalizations of numbers (abstract algebra), moves to shapes (topology), and then adds geometric structure until arriving at fiber bundles. Each chapter attempts to cover the following items:

- Definitions of the main objects under study
- Visualizations of the objects and their relationships
- Related synonyms and jargon common to the literature
- Statements of relevant facts and theorems

Since it is intended partly as a reference, an attempt is made to keep each section of the book as self-contained as possible. The topics are presented in an order designed to minimize any references to future material, and to a large extent this goal is attained. Nevertheless, the book is best read front to back, and results covered in one section are referred to in the sections that immediately follow without further comment.

Notation

Standard notations

The following are standard symbols used in this book. Other symbols will be defined as they are introduced, but the below relations and structures will not be defined and are assumed to already be familiar.

$\forall a \in A, \exists b \mid ab \doteq 0$	For any element a of A, there exists b such that $ab = 0$
$\{x \mid P(x)\}$	The set of elements x that satisfy the relation $P(x)$
$\equiv$	Definition
$=$	Equation derivable from given definitions
$\propto$	Is proportional to
$\subset, \subseteq$	Proper, improper subset, subgroup, etc.
$\cup, \cap$	Union, intersection
$\sum, \prod$	Sum, product
$\binom{n}{k}$	The binomial coefficient n choose k
$\Rightarrow, \Leftrightarrow$	Implies, is true iff (if and only if)
$f : M \to N$	Mapping f from M to N
$m \mapsto n$	Mapping of an element m to an element n
$T\mid_p$	The value of a field T over M at a point $p \in M$
c^*	Complex conjugate of the complex number c
$\mathbb{N}$	The natural numbers
$\mathbb{Z}$	The integers
$\mathbb{Z}^+$	The positive integers
$\mathbb{Z}_n$	The integers modulo n
$\mathbb{R}$	The real numbers
$\mathbb{C}$	The complex numbers

$\mathbb{K}$	A field, typically either the real or complex numbers
$\mathbb{H}$	The quaternions (AKA Hamilton's quaternions)
$\mathbb{O}$	The octonions (AKA Cayley's octonions)
$\mathbb{R}^n$	The n-dimensional real vector space
D^n	The dimension n disk (AKA ball), all vectors of length 1 or less in $\mathbb{R}^n$
S^n	The dimension n sphere, all vectors of length 1 in $\mathbb{R}^{n+1}$
T^n	The dimension n torus, the product of n circles
$A, A^\mu{}_\nu$	Matrix, matrix element of row μ, column ν
$A^T, A^\dagger, I$	Matrix transpose, adjoint, identity
$\det(A), \operatorname{tr}(A)$	Matrix determinant, trace

Defined notations

The following are symbols and conventions used for relations and structures defined in this book. Most are reflective of notation commonly used by other authors, but some are noted as particular to this book.

$\cong$	Isomorphism (between algebraic objects), homeomorphism (between topological spaces), diffeomorphism (between differential manifolds), or isometry (between Riemannian manifolds)
$\simeq$	Homotopy equivalency (between topological spaces)
$\mathbf{1}$	Identity element in a monoid
$\mathbf{0}$	The zero element in a ring
$\operatorname{Ker}, \operatorname{Im}$	The kernel and image of a mapping
$\lvert G \rvert; \lvert g \rvert$	Order of a group G or element g
$\operatorname{Aut}(X)$	The group of all automorphisms of X (group, space, manifold, etc)
$\operatorname{Inn}(G)$	The group of all inner automorphisms of a group G
$N \triangleleft G$	N is a normal subgroup of G
$G = N \rtimes H$	G is the semidirect product of a normal N and H
$\lvert G : H \rvert$	Index of a group G over a subgroup H
G^e	Identity component of a topological group
$V^\perp$	Orthogonal complement of a vector space V
V^*	The dual space of V
$\delta^\mu{}_\nu, \delta_{\mu\nu}$	The Kronecker delta $\equiv 1$ if $\mu = \nu$, 0 otherwise

$\eta^\mu{}_\nu, \eta_{\mu\nu}$	For a pseudo inner product of signature (r,s), ± 1 if $\mu = \nu$ (with r positive values), 0 otherwise
$T^k V$	The kth tensor power of V
$\Lambda^k V$	The kth exterior power of V
M^n	Manifold of dimension n
$\Lambda^k M$	Differential k-forms defined on a manifold M
$T_x M, TM$	Tangent space at x, tangent bundle on M
FM	Frame bundle on M
$\text{Diff}(M)$	The Lie group of diffeomorphisms of a manifold
$\text{vect}(M)$	The Lie algebra of vector fields on a manifold
$\langle v, w \rangle$	Inner product of two vectors
$\|v\|$	Norm
$[u, v]$	Lie bracket
$*A$	Hodge star of $A \in \Lambda^k V$
Ω	Unit n-vector (non-standard)
$\tilde{A}$	Reverse of a Clifford algebra element (in geometric algebra)
$V \times W$	Direct product of two vector spaces
$V \oplus W$	Direct sum
$V * W$	Free product
$V \otimes W$	Tensor product
$V \wedge W$	Exterior product
$\check{\Theta}[\wedge]\check{\Psi}$	Exterior product of Lie algebra valued forms using the Lie commutator (non-standard)
$\check{\Theta} \wedge \check{\Psi}$	Exterior product of Lie algebra valued forms using the multiplication of the related associative algebra
$X \times Y$	Product of two topological spaces
$X \vee Y$	Wedge sum of two spaces
$X * Y$	Join of two spaces
$\mathbb{R}P^n$	Real projective n-space
H^n	Real hyperbolic n-space

Notation conventions

The following are symbols that are typically used to indicate specific types of variables in this book. These conventions will not always be possible to follow, but are as much as is practical.

$G, H; g, h$	Groups; elements of a group
ϕ	Group homomorphism
$R; r$	Ring; element of a ring
$V, W; v, w; a, b$	Vector spaces; elements of a vector space; scalars in a vector space
$\hat{v}$	Unit length vector in a normed vector space
φ, ψ	Mappings or forms
$\vec{\varphi}, \vec{\psi}$	Vector-valued forms (non-standard)
$\check{\Theta}, \check{\Psi}$	Algebra-valued forms (non-standard)
$e_\mu, \hat{e}_\mu$	Basis vectors or frame, orthonormal basis vectors or frame
$\beta^\mu, \hat{\beta}^\mu$	Basis forms or dual frame, orthonormal basis forms or dual frame
$\mathfrak{a}, \mathfrak{b}$	Algebras
$\mathfrak{g}, \mathfrak{h}$	Lie algebras
A, B	Elements of an exterior, Lie, or Clifford algebra
$T^{ab}{}_c$	Tensor using abstract index notation
$T^{\mu_1 \mu_2}{}_{\nu_1}$	Tensor using component notation in a specific basis
X, Y	Topological spaces
M, N	Manifolds
E, P, F	Fiber bundle, principal bundle, abstract fiber

Formatting

A concept is written in **bold** when first mentioned or defined. Also, throughout the text two classes of comments are separated from the core material by boxes:

> $\triangle$ This box will indicate a warning concerning a common confusion or easily misunderstood concept.

> ✿ This box will indicate an interpretation or heuristic view that helps in understanding a particular concept.

Contents

Chapter 1

Mathematical structures

1.1 Classifying mathematical concepts

Most of the mathematical concepts we will be covering will lie in two areas: algebra and geometry. In a very coarse breakdown of the mathematics commonly used in physics, the remaining two broad areas would be foundations and analysis, the basic concepts of which will be presumed to be already familiar to the reader.

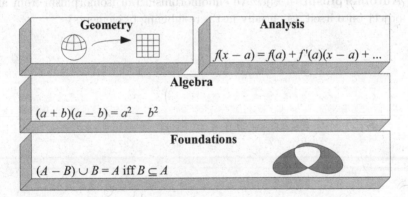

Figure 1.1.1 A coarse breakdown of broad areas in mathematics.

- **Algebra** is principally concerned with sets and arithmetic operations.
- **Geometry** is concerned with spaces and properties that are preserved under various transformations or equivalencies.

In our treatment, we will view algebra as a tool for revealing the structure of geometric objects, which are our primary focus.

1

1.2   Defining mathematical structures and mappings

Most any mathematical object can be viewed as a set of elements along with a "structure." In algebra this structure usually consists of equations that relate the elements to each other, while in geometry the focus is more on relations between subsets of the elements. For example $\mathbb{Z}_2$, the integers modulo 2, is defined by the two elements $\{0, 1\}$ and the equations $0 + 0 = 0, 0 + 1 = 1 + 0 = 1, 1 + 1 = 0$.

We can also introduce the concept of mappings between sets with similar structures. Types of mappings include:

- **Homomorphism**: preserves the structure (e.g. a homomorphism φ on $\mathbb{Z}_2$ satisfies $\varphi(g + h) = \varphi(g) + \varphi(h)$)
- **Epimorphism**: a homomorphism that is **surjective** (AKA onto)
- **Monomorphism**: a homomorphism that is **injective** (AKA one-to-one, 1-1, or univalent)
- **Isomorphism**: a homomorphism that is **bijective** (AKA 1-1 and onto); isomorphic objects are equivalent, but perhaps defined in different ways
- **Endomorphism**: a homomorphism from an object to itself
- **Automorphism**: a bijective endomorphism (an isomorphism from an object onto itself, essentially just a re-labeling of elements)

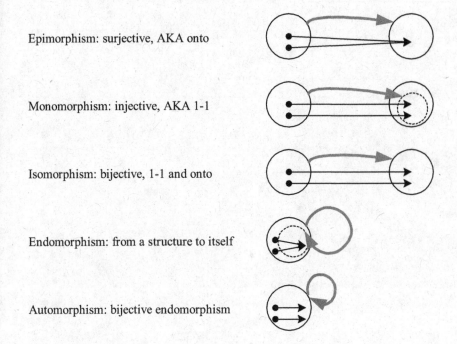

Epimorphism: surjective, AKA onto

Monomorphism: injective, AKA 1-1

Isomorphism: bijective, 1-1 and onto

Endomorphism: from a structure to itself

Automorphism: bijective endomorphism

Figure 1.2.1 Types of mappings between sets.

The concepts of sets and mappings can be generalized further, which takes us from set theory to category theory. An overview of the most basic ideas of category theory is presented in Appendix A. We will occasionally use category theory terminology (object, class, category, morphism and functor) to organize our presentation, but this topic is not necessary to understanding the content of the book.

Chapter 2

Abstract algebra

2.1 Generalizing numbers

Algebra is concerned with sets and operations on these sets. The most common algebraic objects can be viewed as generalizations of the two most familiar examples: the integers and real numbers under addition and multiplication. The generalization starts with a plain set and incrementally adds the properties that define $\mathbb{R}$, yielding objects with increasing structure.

Table 2.1.1 Generalizations of numbers.

	Addition +	Multiplication ×	Special features
Semigroup		Associative	
Monoid		Associative	Unique identity: $1a = a1 = a$
Group		Associative	Inverses: $aa^{-1} = a^{-1}a = 1$
Ring	Abelian group	Semigroup	Zero: $0 + a = a \Rightarrow a0 = 0$
Integral domain	Abelian group	Abelian monoid	No zero divisors
Field	Abelian group	Abelian monoid	Inverses under × except for 0

Notes: $a \times b$ is denoted ab and the identity under × is 1 (other common notations include I and e). For a ring the identity under + is denoted 0 and called zero. Abelian, AKA commutative, means $ab = ba$. The ring operation × is distributive over +. No zero divisors means $ab = 0$ only if $a = 0$ or $b = 0$.

△ It is important to distinguish abstract operations and elements from "ordinary" ones in a particular case. For example, the integers are a

5

group over ordinary addition, so that $a + b$ could be written ab and 0 denoted **1** in a group context. On the other hand, abelian groups are usually written using $+$ as the operation instead of $\times$, with "integer multiplication" defined as $na \equiv a + a + \cdots + a$ (n times); integer multiplication should not be confused with the multiplication of a ring structure, which may be different.

Immediate examples are the real numbers as a field, and the integers as an integral domain; however, the integers are not a group under multiplication since only 1 has an inverse. Some further examples can help illuminate the boundaries between these structures.

- Semigroup but not monoid: the positive reals less than 1 under multiplication (no identity)
- Monoid but not group: the integers under multiplication (no inverses)
- Group but not abelian group: real matrices with non-zero determinant under multiplication
- Abelian group: the integers or real numbers under addition
- Ring but not integral domain: the ring of integers mod n for n not prime (zero divisor $pq = n = 0$)
- Integral domain but not field: the integers (no multiplicative inverses)
- Field: the real numbers; the complex numbers

A **generating set** of an algebraic object is a subset of elements that lead to any other element via operations (e.g. $+$, $\times$). The subset **generates** the object, and the elements in the subset are **generators**. An abelian group is called **finitely generated** if it has a finite generating set.

2.1.1 *Groups*

Groups are one of the simplest and most prevalent algebraic objects in physics. Geometry, which forms the foundation of many physical models, is concerned with spaces and structures that are preserved under transformations of these spaces. At least one source of the prevalence of groups in physics is the fact that if these transformations are automorphisms, they naturally form a group under composition, called the **automorphism group** (AKA symmetry group).

The group of automorphisms of a set is called the **symmetric group**. For the finite set with n elements, the elements of the symmetric group S_n

are called **permutations**, and any subgroup of S_n is called a **permutation group**. The subgroup of all even permutations, i.e. permutations that are obtained by an even number of element exchanges, is called the **alternating group** A_n.

Some common group constructions include:

- **Normalizer** of a subgroup H of G: $N(H) \equiv \{n \in G \mid nHn^{-1} = H\}$
- **Center** of a group: $Z(G) \equiv \{z \in G \mid zg = gz \ \forall g \in G\}$
- **Centralizer** of a subgroup H of G: $C(H) \equiv \{c \in G \mid chc^{-1} = h \ \forall h \in H\}$
- **Inner automorphism** induced by $a \in G$: $\phi_a(g) \equiv aga^{-1}$
- **Order** of an element: $|g|$ is the smallest n such that $g^n = 1$ (may be infinite)
- **Order** of a group: $|G|$ is the number of elements in G (may be infinite)
- **Torsion**: $\text{Tor}(G) \equiv$ elements of finite order; $\text{Tor}(G)$ is a subgroup for abelian G
- **Torsion-free**: $\text{Tor}(G) = 1$

Some of the more important theorems about finite groups include:

- **Cayley's theorem**: every finite group is isomorphic to a group of permutations
- **Lagrange's theorem**: if H is a subgroup of G, $|H|$ divides $|G|$
- **Cauchy's theorem**: if p is a prime that divides $|G|$ then G has an element of order p
- **The fundamental theorem of finite abelian groups**: every finite abelian group can be uniquely written as the direct product of copies of the integers modulo prime powers, with the group operation applied component-wise; i.e. every finite abelian group is of the form

$$\mathbb{Z}_{p_1^{n_1}} \times \mathbb{Z}_{p_2^{n_2}} \times \cdots \times \mathbb{Z}_{p_k^{n_k}}$$

where p_i are not necessarily distinct primes

This last theorem has many consequences, including:

- Any finitely generated abelian group can be written as above, but with some number of $\mathbb{Z}$ components also present
- Any **cyclic group** (generated by a single element) is isomorphic to $\mathbb{Z}_n$
- $|g|$ always divides $|G|$; all groups of prime order are of the form $\mathbb{Z}_n$

We do not discuss normal subgroups here; they will be covered in Section 2.4, "Dividing algebraic objects."

2.1.2   *Rings*

We can define some additional arithmetic generalizations for rings:

- Ring **unity 1** (AKA identity): identity under multiplication; a ring with unity is **unital** (AKA unitary)
- Ring **unit** (AKA invertible element): nonzero element a of commutative ring with multiplicative inverse $aa^{-1} = a^{-1}a = 1$
- **Idempotent** element: element a such that $a^2 = a$
- **Nilpotent** element: there exists an integer n such that $a^n = 0$
- Ring **characteristic**: the least $n \in \mathbb{Z}^+$ such that $na = 0 \ \forall a \in R$; 0 if n does not exist

$\triangle$ It is important to remember that a ring may not have an identity (unity) or inverses under multiplication. However, it should also be noted that "ring" is sometimes defined to include a unity.

As higher structure is added to a ring, it begins to severely constrain its form:

- Every integral domain has characteristic 0 or prime
- Every finite integral domain is a field
- Every finite field is of the form $\mathbb{Z}_{p^n}$, the integers modulo p^n with p prime

We do not discuss ideals here, which are to rings as normal subgroups are to groups, and so are also covered in Section 2.4.

2.2   Generalizing vectors

We can obtain further structure by generalizing the properties of vectors in a Cartesian coordinate system. A **vector space** (AKA linear space) is the algebraic abstraction of the relationships between Cartesian vectors, and it is this structure that we formalize and build up to.

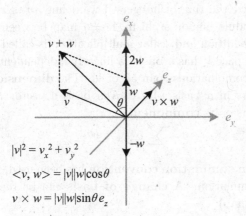

Figure 2.2.1 The structure of Cartesian vectors.

Table 2.2.1 Generalizations of Cartesian vectors.

	Vectors	Scalars	Scalar example
Left R-module	Abelian group V	Ring with unity R	Real matrices
R-module	Abelian group V	Commutative ring R	Integers
Vector space	Abelian group V	Field F	Real numbers
Cartesian space $\mathbb{R}^3$	3-vectors	$\mathbb{R}$	

Notes: The abelian group of vectors has elements denoted u, v, w, with operation $+$ and identity $\mathbf{0}$. The ring of scalars has elements denoted a, b, c, with operations $+$ and $\times$, and special elements $\mathbf{0}$ and $\mathbf{1}$.

For any scalar a and vector v, **scalar multiplication** defines a map to another vector av such that:

- $(ab)v = a(bv)$
- $(a + b)v = av + bv$
- $a(v + w) = av + aw$
- $\mathbf{1}v = v$

A left (right) module defines scalar multiplication only from the left (right), while for the other structures scalar multiplication from either side is equivalent. Modules allow us to generalize real scalars to rings of scalars that lack multiplicative inverses. For example, any abelian group V can be made into a module over the ring of integers if we define $av \equiv v + v + \cdots + v$ (a times). A module homomorphism, i.e. a map between modules that preserves vector addition and scalar multiplication, is called **linear**.

Every vector space V has a **basis**, a linearly independent set of vectors e_μ whose linear combinations span all of V. The **dimension** of V is the number of vectors in a basis. In a given basis a vector v can then be expressed in terms of its **components** v^μ:

$$v = \sum_\mu v^\mu e_\mu \equiv v^\mu e_\mu$$

Here the **Einstein summation convention** has been used, i.e. a repeated index implies summation. A **change of basis** can be represented by a matrix (a linear map):

$$e'_\mu = A^\nu{}_\mu e_\nu$$

Viewed as ordered sequences of vectors, the bases of V can be split into two classes, each class consisting of bases related by a change of basis with positive determinant. A vector space **orientation** is then a choice of one of these two classes.

$\triangle$ It is important to remember that a module may lack a basis or other intuitive features of a vector space.

Over a given field there is a unique vector space (up to isomorphism) of a given dimension; for example $\mathbb{R}^n$ is the only n-dimensional real vector space. In this book we will almost exclusively consider vector spaces over the fields of real or complex numbers. Such vector spaces can be obtained from one another, as follows. The **complexification** of a real vector space, denoted $V_{\mathbb{C}}$, substitutes complex scalars for real ones; thus the basis is left unchanged. The **decomplexification** of a complex vector space W with basis e_μ removes the possibility of complex multiplication of scalars, thus yielding a real vector space $W_{\mathbb{R}}$ of twice the dimension with a basis $\{e_\mu, ie_\mu\}$.

2.2.1 *Inner products of vectors*

For a real or complex vector space V, we can generalize another Cartesian structure, the **inner product** (AKA scalar product, dot product). We define an **inner product space** as including a mapping from vectors to scalars denoted $\langle v, w \rangle$ (also denoted (v, w), $\langle v | w \rangle$, or $v \cdot w$). The mapping must satisfy:

- $\langle v, w \rangle = \langle w, v \rangle^*$ where * denotes complex conjugation
- $\langle ax + y, bz \rangle = a^* b \langle x, z \rangle + b \langle y, z \rangle$
- $\langle v, v \rangle > 0$ except if $v = 0$, in which case it vanishes

The first requirement implies that $\langle v, v \rangle$ is real, and that the inner product is symmetric for real scalars. The second requirement can be phrased as saying that the inner product is **anti-linear** in its first argument and linear in its second, or **sesquilinear**, and the first and second requirements together define a **Hermitian form**. A real product is then **bilinear** or **multilinear**, meaning linear in each argument. The third requirement above makes the inner product **positive definite**.

△ Sometimes the definitions of both inner product and sesquilinear are reversed to make the second argument anti-linear instead of the first. This is sometimes called the "mathematics" convention, while ours would then be the "physics" convention.

Two vectors are defined to be **orthogonal** if their inner product vanishes. The **orthogonal complement** of a subspace W of V is the subspace of all vectors orthogonal to every vector in W, i.e. $W^{\perp} \equiv \{v \in V \mid \forall w \in W \; \langle v, w \rangle = 0\}$. A basis of W together with a basis for its orthogonal complement $W^{\perp}$ forms a basis for all of V.

2.2.2 *Norms of vectors*

In an inner product space, the **norm** (AKA length) of a vector is defined as $\|v\| \equiv \sqrt{\langle v, v \rangle}$, leading to common relations such as the **Cauchy-Schwarz inequality** $|\langle v, w \rangle| \leq \|v\| \, \|w\|$ and the triangle inequality (see below), and letting us define the angle between vectors by $\cos \theta \equiv \langle v, w \rangle / (\|v\| \, \|w\|)$. An inner product then defines a special class of bases, the **orthonormal bases** $\hat{e}_\mu$ with $\langle \hat{e}_\mu, \hat{e}_\nu \rangle = \delta_{\mu\nu}$ ($\equiv 1$ if $\mu = \nu$, 0 otherwise). If we then write $v = v^\mu \hat{e}_\mu$ and $w = w^\mu \hat{e}_\mu$, we have

$$\langle v, w \rangle = \sum_\mu v^{\mu *} w^\mu = v^\dagger w,$$

where in the first expression we take the complex conjugate of the components v^μ, and the second is common in linear algebra, where we treat the vectors as column matrices of components, and the inner product is formed by matrix multiplication after taking the adjoint (hermitian conjugate) of the first matrix.

Alternatively, we can define a **norm** on a real vector space as a non-negative real function that only vanishes for the zero vector and satisfies $\|av\| = |a| \|v\|$ as well as the **triangle inequality** $\|v + w\| \le \|v\| + \|w\|$. This makes the space a real **normed vector space**. Existence of a norm does not in general imply the existence of an inner product, but if a norm satisfies the **parallelogram identity**

$$\|v + w\|^2 + \|v - w\|^2 = 2\left(\|v\|^2 + \|w\|^2\right),$$

an inner product can be obtained using the **polarization identity**

$$\langle v, w \rangle = \left(\|v + w\|^2 - \|v - w\|^2\right)/4 = \left(\|v\|^2 + \|w\|^2 - \|v - w\|^2\right)/2.$$

2.2.3 *Multilinear forms on vectors*

In general, multilinear mappings from a vector space to scalars are called "forms." Here we define some forms with regard to the real vector space $\mathbb{R}^n$; more general definitions may exist in more general settings.

A form is **completely symmetric** if it is invariant under exchange of any two vector arguments; it is **completely anti-symmetric** (AKA alternating) if it changes sign under such an exchange. Some classes of forms include:

- **Bilinear form**: a bilinear map from two vectors to $\mathbb{R}$
- **Multilinear form**: generalizes the above to take any number of vectors
- **Quadratic form**: equivalent to a symmetric bilinear form: a quadratic form is a homogeneous polynomial of degree two, i.e. every term has the same number of variables, with no power greater than 2; by considering the variables components of a vector, the polarization identity gives a 1-1 correspondence with symmetric bilinear forms.
- **Algebraic form**: generalization of the above to arbitrary degree; equivalent to a completely symmetric multilinear form
- **2-form**: an anti-symmetric bilinear form
- **Exterior form**: a completely anti-symmetric multilinear form; exterior forms will be revisited in Section 3.3 from a different perspective.

A real inner product is thus a positive definite quadratic form, and is the generalization to arbitrary dimension of the geometrically defined dot product in $\mathbb{R}^3$. In the context of manifolds (see Chapter 6), an inner product is called a **metric**; and in the context of spacetime, variants of the inner product carry specific terminology. We will cover this terminology in the following.

The positive definite definition of a real inner product is sometimes relaxed to only require the product to be **nondegenerate** (AKA anisotropic), i.e. $\langle v, w \rangle = 0$ for all w only if $v = 0$. We will instead refer to this type of form, a nondegenerate symmetric bilinear form, as a **pseudo inner product** (AKA pseudo-metric); it is characterized by the fact that $\langle v, v \rangle$ can be negative or vanish. A vector v is called **isotropic** (AKA light-like) if $\langle v, v \rangle = 0$.

A pseudo inner product does not yield a well-defined norm, but the (not necessarily real) quantity $\sqrt{\langle v, v \rangle}$ is nevertheless sometimes called the "length" of v. A pseudo inner product also defines orthonormal bases, with the definition modified to allow $\langle \hat{e}_\mu, \hat{e}_\nu \rangle = \pm\delta_{\mu\nu}$. The number of positive and negative "lengths" of basis vectors turns out to be independent of the choice of basis, and this pair of integers (r, s) is called the **signature**. We define $\eta_{\mu\nu} \equiv \langle \hat{e}_\mu, \hat{e}_\nu \rangle \equiv \eta^{\mu\nu}$, so that it is ± 1 if $\mu = \nu$ (with r positive values), 0 otherwise. A general signature is called **pseudo-Riemannian** (AKA pseudo-Euclidean), a signature with $s = 0$ is called **Riemannian** (AKA Euclidean), a signature with $s = 1$ (or sometimes $r = 1$) is called **Lorentzian** (AKA Minkowskian), and the signature (3,1) is called **Minkowskian** (more specifically this is called the **"mostly pluses" signature** (AKA relativity, spacelike, or east coast signature) while the signature (1,3), the **"mostly minuses" signature** (AKA particle physics, timelike, or west coast signature), is also called Minkowskian).

$\triangle$ The term "signature" also sometimes refers to the integer $r - s$.

Below we summarize the relationships between various forms.

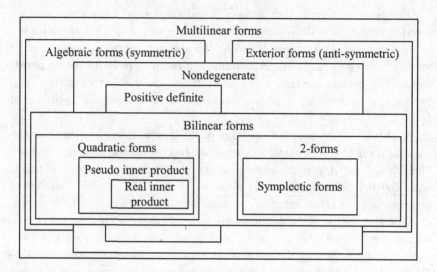

Figure 2.2.2 Various types of forms on $\mathbb{R}^n$.

✿ The above is an **Euler diagram**, where the spatial relationships of the boxes indicate their relationships as sets (i.e. intersection, subset, disjoint). We will use these frequently in summarizing the relationships between mathematical concepts.

2.2.4 *Orthogonality of vectors*

Geometrically, one can alternatively take the approach that "orthogonality" is the fundamental structure to be added to a real vector space $V = \mathbb{R}^n$. Orthogonality is defined by requiring that:

(1) Any subspace W defines an orthogonal complement $W^\perp$ such that only the zero vector is contained in both spaces (an **orthogonal decomposition**)

(2) If v is orthogonal to w, then w is orthogonal to v

One can then look for bilinear forms that vanish for orthogonal vector arguments. A degenerate form features at least one vector v that is orthogonal to every other vector in V, thus violating (1) by being in both V and $V^\perp$. By linearity, (2) can only be satisfied if the form is either symmetric or anti-symmetric. Thus our only two candidates are a pseudo inner product or a nondegenerate 2-form. The latter is also called a **symplectic form**, and can only exist in even-dimensional vector spaces.

2.2.5 *Algebras: multiplication of vectors*

The abelian group of vectors in a vector space can also be given a new structure by defining multiplication between vectors to get another vector. The Cartesian inspiration here might be considered to be the vector **cross product** (AKA vector product or outer product) in 3 dimensions.

- **Algebra**: defines a bilinear product distributive over vector addition (no commutativity, associativity, or identity required)
- **Associative algebra**: associative product; turns the abelian group of vectors into a ring; can always be naturally extended to include a multiplicative identity
- **Lie algebra**: product denoted $[u,v]$; satisfies two other attributes of the cross product, anti-commutativity $[u,v] = -[v,u]$ and the **Jacobi identity** $[[u,v],w] + [[w,u],v] + [[v,w],u] = 0$

For example, the Cartesian vectors under the cross product are a non-associative Lie algebra, while the real $n \times n$ matrices under matrix multiplication are an associative algebra.

$\triangle$ Note that "algebra" is sometimes defined to include associativity and/or an identity.

$\triangle$ If the scalars of a Lie algebra are a field of characteristic 2, then we no longer have $[u,v] = -[v,u] \Rightarrow [v,v] = 0$, and the latter is imposed as a separate requirement in the definition.

In a Lie algebra (pronounced "lee"), the product is called the **Lie bracket**, and the notation $[u,v]$ in place of uv reflects the close relationship between Lie algebras and associative algebras: every associative algebra can be turned into a Lie algebra by defining the Lie bracket to be $[u,v] \equiv uv - vu$. In these cases the Lie bracket is called the **Lie commutator**. The **Poincaré-Birkhoff-Witt theorem** provides a converse to this: that every Lie algebra is isomorphic to a subalgebra of an associative

algebra called the **universal enveloping algebra** under the Lie commutator. An **abelian algebra** has $uv = vu$; thus an **abelian Lie algebra**
(AKA commutative Lie algebra) has $[u, v] = [v, u] \Rightarrow [u, v] = 0$. Note
that an associative Lie algebra is not necessarily abelian, but does satisfy
$[[u, v], w] = 0$ via the Jacobi identity.

Any algebra over a field is completely determined by specifying scalars
called **structure coefficients** (AKA structure constants), defined in a
given basis as follows:

$$e_\mu e_\nu = c^\rho{}_{\mu\nu} e_\rho$$

However, different structure coefficients may define isomorphic algebras.

2.2.6 *Division algebras*

A **division algebra** is an algebra with a multiplicative identity where
unique right and left inverses exist for every non-zero element. For an
associative division algebra, these inverses are equal, turning the non-zero
vectors into a group under multiplication. Ignoring scalar multiplication, an
associative algebra is a ring, and a commutative associative division algebra
is a field; thus an associative division algebra with scalar multiplication
ignored is sometimes called a **division ring** or **skew field**. A module over
a non-commutative skew field (such as $\mathbb{H}$) can be seen to have much of the
same features as a vector space, including a basis.

In a division algebra, the existence of the left inverse u_L^{-1} of u allows
us to "divide" elements in the sense that for any non-zero u and v, $xu = v$
has the solution $x = vu_L^{-1}$, which we can regard as the "left" version of
v/u; similarly, $ux = v$ has the solution $x = u_R^{-1}v$. This is equivalent to
requiring that there be no **zero divisors**, i.e. $uv = 0 \Rightarrow u = 0$ or $v = 0$.
A **normed division algebra** has a vector space norm that additionally
satisfies $\|uv\| = \|u\| \, \|v\|$.

Finite-dimensional real division algebras are highly constrained: all have
dimension 1, 2, 4, or 8; the commutative ones all have dimension 1 or 2;
the only associative ones are $\mathbb{R}$, $\mathbb{C}$, and $\mathbb{H}$; and the only normed ones are
$\mathbb{R}$, $\mathbb{C}$, $\mathbb{H}$, and $\mathbb{O}$. Here we review these division algebras:

- $\mathbb{C}$, the complex numbers, has basis $\{1, i\}$ where $i^2 \equiv -1$
- $\mathbb{H}$, the **quaternions**, has basis $\{1, i, j, k\}$ where $i^2 = j^2 = k^2 = ijk \equiv -1$
- $\mathbb{O}$, the **octonions**, has basis $\{1, i, j, k, l, li, lj, lk\}$, all anti-commuting
  square roots of -1; we will not describe the full multiplication table
  here

We can define the **quaternionic conjugate** by reversing the sign of the i, j, and k components, with the **octonionic conjugate** defined similarly. The norm is then defined by $\|v\| = \sqrt{vv^*}$ in these algebras, as it is in $\mathbb{C}$.

We lose a property of the real numbers each time we increase dimension in the above algebras: $\mathbb{C}$ is not ordered; $\mathbb{H}$ is not commutative; and $\mathbb{O}$ is not associative. $\mathbb{C}$ is a field (ignoring real scalar multiplication) and so can be used as the scalars in a vector space $\mathbb{C}^n$. One could imagine then trying to find a multiplication on $\mathbb{C}^n$ to obtain complex division algebras, but the only finite-dimensional complex division algebra is $\mathbb{C}$ itself. The quaternions form a non-commutative ring, and so can be used as the scalars in a left module $\mathbb{H}^n$, but there is no obvious definition of $\mathbb{O}^n$ since the octonions are not associative. However, we can form all of the algebras $\mathbb{R}(n)$, $\mathbb{C}(n)$, $\mathbb{H}(n)$, and $\mathbb{O}(n)$, where $\mathbb{K}(n)$ denotes the algebra of $n \times n$ matrices with entries in $\mathbb{K}$. $\mathbb{H}(n)$ can even be viewed as the group of linear transformations on $\mathbb{H}^n$, if $\mathbb{H}^n$ is defined as a right module while matrix multiplication takes place from the left as usual.

2.3 Combining algebraic objects

We can define combinations of algebraic objects to construct new, "bigger" objects in the same category. We will use the concepts of categorical **products** and **coproducts** (AKA sums) in category theory to organize our presentation. While we will not go into exact definitions here, a categorical product can be thought of as the "most general" object with morphisms to its constituents, while a categorical coproduct can be thought of as the "most general" object with morphisms in the opposite direction, from the constituent objects to their coproduct. In certain categories, the product and coproduct of two objects coincide, in which case they are both called the **biproduct** (AKA direct sum). Even in these categories, however, the product and coproduct are distinct in the case of an infinite number of factors. Note that the common meaning of "direct sum" is not equivalent to the categorical direct sum (biproduct) in category theory, as we see below.

Table 2.3.1 The categorical product and coproduct in different categories.

	Product	Coproduct
Sets	Cartesian product $A \times B$	Disjoint union $A \cup_d B$
Groups	Direct product $G \times H$	Free product $G * H$
Abelian groups	Direct product $G \times H$	Direct sum $G \oplus H$
Vector spaces	Direct product $V \times W$	Direct sum $V \oplus W$
Commutative rings with unity	Direct product $R \times S$	Tensor product $R \otimes S$

Notes: Coproducts in algebras and other categories can become quite complicated.

2.3.1   *The direct product and direct sum*

The **direct product** takes the **Cartesian product** $A \times B$ of sets, i.e. the ordered pairs of elements (a, b), and applies all operations component-wise; e.g. for a group we define $(a, b) + (c, d) \equiv (a + c, b + d)$. Note that this approach cannot be taken in all categories; for example, a new field cannot be obtained from the direct product of two fields, since $(0, a)$ is distinct from $\mathbf{0}$ and has no multiplicative inverse. The **direct sum** is identical to the direct product except in the case of an infinite number of factors, when the direct sum $\bigoplus A_\mu$ consists of elements that have only finitely many non-identity terms, while the direct product $\prod A_\mu$ has no such restriction.

$\triangle$ For a finite number of objects, the direct product and direct sum are identical constructions, and these terms are often used interchangeably, along with their symbols $\times$ and $\oplus$. In particular, since the group operation is usually written like multiplication, we usually write $G \times H$; with vector spaces and algebras, where the abelian group operation is written like addition, we instead write $V \oplus W$.

$\triangle$ The categorical coproduct for abelian groups and vector spaces is the direct sum, which is then also applied to objects in other categories, potentially causing confusion with the "categorical direct sum" (biproduct) in those other categories, which may be a distinct construction or not exist.

An additional distinction that can be made is between an **external** sum or product, where the construction of a new object is from given constituent objects, and an **internal** sum or product, which is formed from a given object by recognizing constituent sub-objects within it and noting that the sum or product of the sub-objects is isomorphic to the original object.

> △ In addition to denoting a direct sum, the symbol $\oplus$ is sometimes used to denote either the internal direct product of groups or the free product. In general, the above symbols are not always used consistently and it is important to understand exactly what operation is meant in a given situation.

As we will see, in addition to this "internal/external" distinction, there is a rough logic as well to the distinction between the designations "inner/outer" and "interior/exterior" with regard to products.

2.3.2 The free product

The coproduct in the category of groups is the **free product** $G * H$, defined as the set of finite ordered "words" $g_1 h_2 g_3 h_4 \cdots g_{n-1} h_n$ of non-identity elements, with the group operation applied as juxtaposition of words. After juxtaposition, any adjacent letters from the same group are combined, and removed if the result is the identity. Consistent with this, the free product of a family of groups G_μ, denoted $*G_\mu$, is defined as finite words with the letters being non-identity elements from any G_μ, where no adjacent elements come from the same G_μ.

> △ It is important to note the difference between the free product $*G_\mu$, where each letter g_i is an element of any G_μ distinct from the previous one, and the direct sum $\bigoplus G_\mu$, which can be viewed as a word where each letter g_μ is an element of a distinct G_μ (i.e. g_μ is the only element of G_μ in the word).

The free product is an example of the more general **free object** in category theory, which can be thought of as "forcing" one category into being another in the "most general" way; again we will not go into exact definitions, but instead describe some common free constructions.

- The **free group** on a set S "forces" S into being a group, defining inverses by a copy S' and forming a group out of the finite ordered words of elements of $S \cup S'$ with juxtaposition as the group operation (as with the free product, any combinations ss^{-1} or $s^{-1}s$ are removed)
- The **free associative algebra** on a vector space V (AKA the associative algebra W freely generated by V) "forces" the words $v_1 v_2 \cdots v_n$ into being an associative algebra by defining vector multiplication as juxtaposition and requiring it to be multilinear, i.e. $(v_1 + v_2)(av_3) \equiv av_1v_3 + av_2v_3$; as we will see below, this is in fact just the tensor algebra over V, so an element $v_1 v_2$ can be written $v_1 \otimes v_2$
- The **free module** of rank n over a ring R has no multiplication, so the words of a specified length $r_1 r_2 \cdots r_n$ are "forced" into being a module by defining addition and multiplication component-wise, i.e.

$$t(r_1 r_2 \cdots r_n + s_1 s_2 \cdots s_n) = (tr_1 + ts_1)(tr_2 + ts_2) \cdots (tr_n + ts_n);$$

thus an element $r_1 r_2$ is just a direct sum, and can be written $r_1 \oplus r_2$

The **free abelian group** is the free module of rank n over $\mathbb{Z}$, since as we noted previously, any abelian group can be viewed as a module over $\mathbb{Z}$ under "integer multiplication." In fact, the free abelian group of rank n is just $\mathbb{Z} \oplus \mathbb{Z} \oplus \cdots \oplus \mathbb{Z}$ (n times) under component-wise addition.

$\triangle$ Note that the name "free abelian group" is a potential source of confusion, since it is a free module, not a free group (except for the case of rank one, i.e. $\mathbb{Z}$).

2.3.3 *The tensor product*

The tensor product appears as a coproduct for commutative rings with unity, but as with the direct sum this definition is then extended to other categories. For abelian groups, the **tensor product** $G \otimes H$ is the group generated by the ordered pairs $g \otimes h$ linear over $+$; as more structure is added, the tensor product is required to be bilinear with regard to these structures. It can then be applied to multiple objects by extending these bilinear rules to multilinear ones.

It is helpful to compare the properties of the tensor product to the direct sum in various categories, since consistent with their symbols $\oplus$ and $\otimes$ they act in many ways like addition and multiplication.

Table 2.3.2 The direct sum and tensor product in different categories.

	Direct sum $\oplus$	Tensor product $\otimes$
Abelian groups	$(v_1 \oplus w) + (v_2 \oplus w)$	$(v_1 \otimes w) + (v_2 \otimes w)$
	$\equiv (v_1 + v_2) \oplus (w + w)$	$\equiv (v_1 + v_2) \otimes w$
Vector spaces	$a(v \oplus w)$	$a(v \otimes w)$
	$\equiv av \oplus aw$	$\equiv av \otimes w \equiv v \otimes aw$
Inner product spaces	$\langle v_1 \oplus w_1, v_2 \oplus w_2 \rangle$	$\langle v_1 \otimes w_1, v_2 \otimes w_2 \rangle$
	$\equiv \langle v_1, v_2 \rangle + \langle w_1, w_2 \rangle$	$\equiv \langle v_1, v_2 \rangle \langle w_1, w_2 \rangle$
Algebras / Rings	$(v_1 \oplus w_1)(v_2 \oplus w_2)$	$(v_1 \otimes w_1)(v_2 \otimes w_2)$
	$\equiv (v_1 v_2) \oplus (w_1 w_2)$	$\equiv (v_1 v_2) \otimes (w_1 w_2)$

Notes: The addition and multiplication of inner products is that of scalars, while the multiplication of vectors is that of the algebra or ring.

It is important to remember that elements of the direct sum $V \oplus W$ always have the form $v \oplus w$, while elements of the tensor product $V \otimes W$ are generated by the elements $v \otimes w$ using the operation $+$ as defined above, so that the general element of $V \otimes W$ has the form of a sum $\sum (v_\mu \otimes w_\nu)$. For example, if V and W are m- and n-dimensional vector spaces with bases d_μ and e_ν, $V \otimes W$ has basis $\{d_\mu \otimes e_\nu\}$ and dimension mn, while $V \oplus W$ has basis $\{d_1, \ldots, d_m, e_1, \ldots, e_n\}$ and dimension $m + n$. If V and W are algebras defined by square matrices, the direct sum $V \oplus W$ and tensor product $V \otimes W$ have elements that are isomorphic to matrices that can be formed from the matrices v and w:

$$v \oplus w \cong \begin{pmatrix} v & 0 \\ 0 & w \end{pmatrix} \qquad v \otimes w \cong \begin{pmatrix} v_{11}w & v_{12}w & \cdots \\ v_{21}w & v_{22}w & \cdots \\ \vdots & \vdots & \ddots \end{pmatrix}$$

The convention used in the second isomorphism, in which $v \otimes w$ is "the matrix v with elements multiples of w," is sometimes called the **Kronecker product**; one can also choose to use the opposite convention. Some other specific isomorphisms include:

- $\mathbb{C} \otimes \mathbb{C} \cong \mathbb{C} \oplus \mathbb{C}$
- $\mathbb{C} \otimes \mathbb{H} \cong \mathbb{C}(2)$
- $\mathbb{H} \otimes \mathbb{H} \cong \mathbb{R}(4)$

where e.g. $\mathbb{C}(2)$ denotes the algebra of complex 2×2 matrices.

2.4   Dividing algebraic objects

Above, we combined objects using sums and products to form larger objects. Here, this generalization of arithmetic is extended to quotients, starting with groups.

2.4.1   *Quotient groups*

The first step in this direction is to define cosets. For any subgroup $H \subset G$ and element $g \in G$, the set $gH \equiv \{gh \mid h \in H\}$ is called a left **coset** of H in G. Right cosets are defined similarly, and g is called a **representative element** of the coset gH. Cosets can be viewed as a partitioning of all of G into equal-sized disjoint "copies." However, this cannot be used to define a group quotient G/H since in general, the cosets themselves do not form a group.

A **normal subgroup** (AKA invariant subgroup, self-conjugate subgroup) of G, denoted $N \triangleleft G$, is defined as follows:

$$N \triangleleft G \text{ if } gN = Ng \; \forall g \in G \Leftrightarrow gNg^{-1} \subseteq N \; \forall g \in G$$

Note that an immediate consequence of the above definition is that any subgroup of an abelian group is normal. It is not hard to see that the cosets of N in G (left and right being identical) do in fact comprise the elements of a group under the group operation $(gN)(hN) \equiv \{gnhm \mid n, m \in N\} = (gh)N$. We denote this group G/N and call it a **quotient group** (AKA factor group).

The **kernel** Kerϕ of a homomorphism ϕ from G to another group Q is the subgroup of elements that are mapped to the identity **1**. Any normal subgroup N is then the kernel of the group homomorphism $\phi \colon G \to G/N$ defined by $g \mapsto gN$, and thus all normal subgroups are homomorphism kernels. The converse of this is also true: for any homomorphism $\phi \colon G \to Q$, Kerϕ is normal in G. Furthermore, the quotient group is isomorphic to the subgroup $\phi(G)$ of Q, so that we have the equation $G/\text{Ker}\phi \cong \phi(G)$, called the **first isomorphism theorem** or the **fundamental theorem on homomorphisms**: ϕ shrinks each equal-sized coset of G to an element of $\phi(G)$, which is therefore a kind of simpler approximation to G.

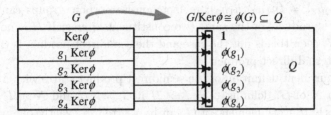

Figure 2.4.1 The kernel of a homomorphism $\phi\colon G \to Q$ factors the group into elements of the image.

It is helpful to demonstrate quotient groups with an easy example. Let $\phi\colon \mathbb{Z} \to \mathbb{Z}_3$ be the (surjective) homomorphism that sends each element to its remainder after being divided by 3. The kernel of this homomorphism is the subgroup of $\mathbb{Z}$ consisting of all integers of the form $3n$. Then the cosets $\mathbb{Z}/\{3n\}$ are the subgroups $\{3n\}$, $\{3n+1\}$, and $\{3n+2\}$, which are isomorphic to the elements of $\mathbb{Z}_3$.

2.4.2 Semidirect products

In general, $\phi(G) \cong G/\mathrm{Ker}\phi$ is not a subgroup of G, so we cannot use this equation to decompose G into an internal product by "multiplying both sides by $\mathrm{Ker}\phi$." If H is a subgroup of G and there is a homomorphism $\phi\colon G \to H$ that is the identity on H, we can once again extend the arithmetic of groups to state that $G = \mathrm{Ker}\phi \rtimes H$, where $\rtimes$ represents the **semidirect product** (sometimes stated "G splits over $\mathrm{Ker}\phi$").

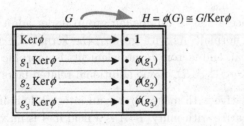

Figure 2.4.2 A homomorphism from G to a subgroup H that is the identity on H induces a semidirect product structure on G.

Note that the semidirect product as defined here is an internal product formed from G, and is distinct from the direct product since in general

$(n_1h_1)(n_2h_2) \neq (n_1n_2)(h_1h_2)$; in fact non-isomorphic groups can be the semidirect products of the same two constituent groups. If H is normal as well as N, then this is not the case and the semidirect product is the same as the internal direct product.

An equivalent definition of the semidirect product starts with a normal subgroup N of G, defining $G = N \rtimes H$ if $G = NH$ and $N \cap H = 1$, or equivalently if every element of G can be written in exactly one way as a product of an element of N and an element of H. These properties can be seen in a common use of the semidirect product in physics, where N is the group of translations in $\mathbb{R}^n$, H is the group of rotations and reflections, and $N \rtimes H$ is thus the group of all rigid transformations.

2.4.3 Quotient rings

We can also define the **quotient ring** (AKA factor ring) of a ring R, and related concepts:

- **Ideal**: subring $A \subset R$ where $ra, ar \in A \; \forall a \in A, r \in R$
- **Quotient ring**: The cosets $R/A \equiv \{r + A \mid r \in R\}$, which form a ring iff A is an ideal
- **Prime ideal**: Subring $A \subset R \mid ab \in A$ for $a, b \in R \Rightarrow a \in A$ or $b \in A$
- **Maximal ideal**: $\forall$ ideal $B \supseteq A$, $B = A$ or $B = R$

The definition of ideal above is sometimes called a **two-sided ideal**, in which case a **left ideal** only requires that $ra \in A$ and a **right ideal** requires that $ar \in A$. For a commutative ring, these are all equivalent. These concepts are also applied to associative algebras, since with scalars ignored they are rings.

Note that since a ring is an abelian group under addition, every subgroup is already normal. As with groups, the kernel of a ring homomorphism ϕ is an ideal, and factors R into elements isomorphic to those of the image of R: $R/\text{Ker}\phi \cong \phi(R)$. Some additional related facts are:

- For R commutative with unity, R/A is an integral domain iff A is prime
- For R commutative with unity, R/A is a field if A is maximal

Continuing to add structure, in a vector space V we can take the quotient V/W for any subspace W, which is just isomorphic to the orthogonal complement of W in V.

2.4.4   Related constructions and facts

Another important class of groups is the **simple groups**, whose only normal subgroups are G and **1**. The finite simple groups behave in many ways like primes, and after a long effort have been fully classified. Some additional constructions and facts concerning quotients include:

- The **index** of a group G over a subgroup H, denoted $|G : H|$, is the number of cosets of H in G
- $|G : H| = |G|/|H|$ for finite groups
- $G/Z(G) \cong \mathrm{Inn}(G)$, the group of all inner automorphisms of G
- $G/Z(G)$ is cyclic, G is abelian
- The abelian simple groups are $\mathbb{Z}_p$ for p prime
- **Feit-Thompson theorem**: non-abelian simple groups have even order

2.5   Summary

In summary, it is helpful to see how algebraic structures follow by discarding properties of the real numbers, and then how more complicated structures are constructed by generalizing vectors and scalars.

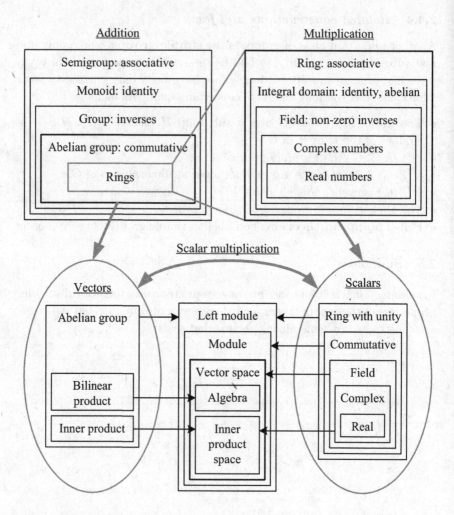

Figure 2.5.1 Summary of algebraic categories and their relationships.

Chapter 3

Vector algebras

3.1 Constructing algebras from a vector space

When applied to two copies of the same vector space V, the tensor product is sometimes called the "outer product," since it is a linear map from two vectors to a "bigger" object "outside" V, as opposed to the inner product, which is a linear map from two vectors to a "smaller" object "inside" V.

The term "outer product" also sometimes refers to the **exterior product** $V \wedge V$ (AKA wedge product, Grassmann product), which is defined to be the tensor product $V \otimes V$ modulo the relation $v \wedge v \equiv 0$. By considering the quantity $(v_1 + v_2) \wedge (v_1 + v_2)$, this immediately leads to the equivalent requirement of identifying anti-symmetric elements, $v_1 \wedge v_2 \equiv -v_2 \wedge v_1$.

These outer products can be applied to a vector space V to generate a larger vector space, which is then an algebra under the outer product. If V has a pseudo inner product defined on it, we can then generalize it to the larger algebra. In the following, we will limit our discussion to finite-dimensional real vector spaces $V = \mathbb{R}^n$; generalization to complex scalars is straightforward.

3.1.1 *The tensor algebra*

We first use the tensor product to generate an algebra from an n-dimensional vector space V. The kth **tensor power** of V, denoted $T^k V$, is the tensor product of V with itself k times; it forms a vector space of dimension n^k. The **tensor algebra** of V is then defined to be the infinite direct sum of every tensor power:

$$TV \equiv \sum T^k V = \mathbb{R} \oplus V \oplus (V \otimes V) \oplus (V \otimes V \otimes V) \oplus \cdots$$

(however, see the last paragraph of Section 3.2.2 on page 40).

The vector multiplication operation is $\otimes$, and thus the infinite-dimensional tensor algebra is associative. In fact, the tensor algebra can alternatively be defined as the free associative algebra on V, with juxtaposition indicated by the tensor product.

If a pseudo inner product, i.e. a nondegenerate symmetric bilinear form, is defined on V, it can be naturally extended to any $T^k V$ by extending the pairwise operation defined previously: if $A = v_1 \otimes v_2 \otimes \cdots \otimes v_k$, and $B = w_1 \otimes w_2 \otimes \cdots \otimes w_k$, we define

$$\langle A, B \rangle \equiv \prod \langle v_i, w_i \rangle.$$

The pseudo inner product can then be extended to a nondegenerate symmetric multilinear form on all of TV by defining it to be zero between elements from different tensor powers.

3.1.2 The exterior algebra

Similarly, the kth **exterior power** of an n-dimensional vector space V is defined to be

$$\Lambda^k V \equiv V \wedge V \wedge \cdots \wedge V \qquad (k \text{ times}).$$

The exterior product is generalized to $\Lambda^k V$ by requiring the product to vanish if any two vector components are identical. This is equivalent to requiring the product to be completely anti-symmetric, i.e. to change sign under the exchange of any two vector components. Note that $\Lambda^k V$ thus automatically vanishes for $k > n$, since the $(k+1)^{\text{th}}$ component will have to be a linear combination of previous components, resulting in terms $v \wedge v = 0$.

The **exterior algebra** (AKA Grassmann algebra, alternating algebra) is the tensor algebra modulo the relation $v \wedge v \equiv 0$, and can be written as

$$\Lambda V \equiv \Sigma \Lambda^k V = \mathbb{R} \oplus \Lambda^1 V \oplus \Lambda^2 V \oplus \cdots \oplus \Lambda^n V,$$

where n is the dimension of V (since $\Lambda^k V$ automatically vanishes for $k > n$). The vector multiplication of the algebra is the exterior product $\wedge$, which applied to $A \in \Lambda^j V$ and $B \in \Lambda^k V$ gives $A \wedge B \in \Lambda^{j+k} V$ with the property

$$A \wedge B \equiv (-1)^{jk} B \wedge A.$$

If a pseudo inner product is defined on V, it can be naturally extended to any $\Lambda^k V$ by using the determinant: if $A = v_1 \wedge v_2 \wedge \cdots \wedge v_k$, and $B = w_1 \wedge w_2 \wedge \cdots \wedge w_k$, we define

$$\langle A, B \rangle \equiv \det \left(\langle v_i, w_j \rangle \right).$$

Note that this definition is also alternating, as it must for the inner product to be bilinear; i.e. exchanging two vectors in A reverses the sign of both A and $\langle A, B \rangle$. Also note that if $A = ae_1 \wedge \cdots \wedge e_k$, then $\langle A, A \rangle = \pm a^2$. The pseudo inner product can then be extended to a nondegenerate symmetric multilinear alternating form on all of ΛV by defining it to be zero between elements from different exterior powers. If $\hat{e}_\mu$ is an orthonormal basis for V, then

$$\{\hat{e}_{\mu_1} \wedge \cdots \wedge \hat{e}_{\mu_k}\}_{1 \leq \mu_1 < \cdots < \mu_k \leq n}$$

is an orthonormal basis for $\Lambda^k V$, and the union of such bases for $k \leq n$ is an orthonormal basis for ΛV.

In describing a particular element of ΛV we can unambiguously write $+$ instead of $\oplus$ and allow any zero terms to be omitted. A simple example can be helpful to keep in mind the concrete consequences of the exterior algebra's abstract properties.

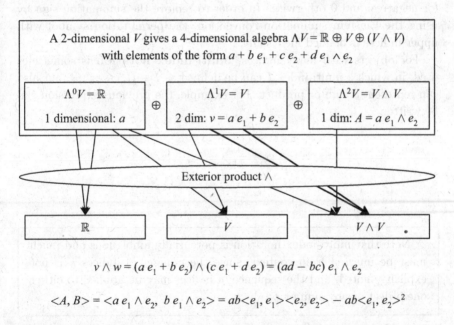

Figure 3.1.1 The exterior algebra over a real 2-dimensional vector space. Exterior products not shown above vanish; the product indicated with bold arrows is elaborated by the first equation. The second equation calculates the inner product of two elements of $\Lambda^2 V$ using the determinant.

3.1.3 *Combinatorial notations*

An alternating quantity can be represented in several different ways. For example, the exterior product applied to multiple vectors is defined to change sign under the exchange of any two vector components. This can be written

$$v_1 \wedge v_2 \wedge \cdots \wedge v_k = \text{sign}(\pi) v_{\pi(1)} \wedge v_{\pi(2)} \wedge \cdots \wedge v_{\pi(k)},$$

where π is any permutation of the k indices, and $\text{sign}(\pi)$ is the sign of the permutation. Another way of writing it is

$$v_1 \wedge v_2 \wedge \cdots \wedge v_k = \frac{1}{k!} \sum_{i_1, i_2, \ldots, i_k} \varepsilon_{i_1 i_2 \ldots i_k} v_{i_1} \wedge v_{i_2} \wedge \cdots \wedge v_{i_k},$$

where each index ranges from 1 to k and ε is the **permutation symbol** (AKA completely anti-symmetric symbol, Levi-Civita symbol, alternating symbol, ε-symbol), defined to be $+1$ for positive index permutations, -1 for negative, and 0 otherwise. In order to remove the summation sign by using the Einstein summation convention, the permutation symbol with upper indices is defined identically.

For objects with many indices, **multi-index notation** is sometimes used, in which a multi-index I can be defined as $I \equiv i_1, i_2, \ldots, i_k$, but also can represent a sum or product. For example, the previous expression can be written

$$v_1 \wedge v_2 \wedge \cdots \wedge v_k = \frac{1}{k!} \sum_I \varepsilon^I v_{i_1} \wedge v_{i_2} \wedge \cdots \wedge v_{i_k}$$

$$= \frac{1}{k!} \varepsilon^I v_I.$$

$\triangle$ Note that multi-index notation is potentially ambiguous and much must be inferred from context, since the number of indices k is not explicitly noted, and the sequence of indices may be applied to either one object or any sum or product.

Another example of an alternating quantity is the determinant of an $n \times n$ matrix $M^i{}_j$, which can be written

$$\det(M) = \sum_{\pi} \text{sign}\,(\pi)\, M^1{}_{\pi(1)} M^2{}_{\pi(2)} \cdots M^n{}_{\pi(n)}$$

$$= \sum_{i_1,i_2,\ldots,i_n} \varepsilon^{i_1 i_2 \ldots i_n} M^1{}_{i_1} M^2{}_{i_2} \cdots M^n{}_{i_n},$$

where the first sum is over all permutations π of the n second indices of the matrix $M^i{}_j$. Using the previous relation for the exterior product in terms of the permutation symbol, we can see that the transformation of the top exterior product of basis vectors under a change of basis $e'_\mu = M^\nu{}_\mu e_\nu$ is

$$e'_1 \wedge e'_2 \wedge \cdots \wedge e'_n = \det(M)\, e_1 \wedge e_2 \wedge \cdots \wedge e_n,$$

which reminds of us of the Jacobian determinant from integral calculus, and as we will see makes the exterior product a natural way to express the volume element.

☆ The above relationship between the exterior product and the determinant means that under a positive definite inner product, for k arbitrary vectors $v_\mu = M^\nu{}_\mu e_\nu$ the quantity $P \equiv v_1 \wedge v_2 \wedge \cdots \wedge v_k$ satisfies $\sqrt{\langle P, P \rangle} = \det(M)$, which is the volume of the parallelepiped defined by the vectors. There are other ways in which P behaves like a parallelepiped, and it is often useful to picture it as such.

If V is n-dimensional and has a basis e_μ, a general element A of $\Lambda^k V$ can be written in terms of a basis for $\Lambda^k V$ as

$$A = \sum_{\mu_1 < \cdots < \mu_k} A^{\mu_1 \cdots \mu_k} e_{\mu_1} \wedge \cdots \wedge e_{\mu_k}.$$

Here the sum is over only ordered sequences of indices, since due to anti-symmetric elements being identified, only these are linearly independent. Each index can take on any value between 1 and n. We can also write

$$A = \frac{1}{k!} \sum_{\mu_1,\ldots,\mu_k} A^{\mu_1 \cdots \mu_k} e_{\mu_1} \wedge \cdots \wedge e_{\mu_k},$$

where the coefficient is now defined for all combinations of indices, and its value changes sign for any exchange of indices (and thus vanishes if any two indices have the same value). The factorial ensures that the values for ordered sequences of indices matches the above expression.

The first expression shows that $\Lambda^k V$ is a vector space with dimension equal to the number of distinct subsets of k indices from the set of n available, i.e. its dimension is equal to the binomial coefficient "n choose k"

$$\binom{n}{k} \equiv \frac{n!}{k!\,(n-k)!}.$$

A general element of ΛV then has the form

$$\bigoplus_{0 \leq k \leq n} \left[\sum_{\mu_1 < \cdots < \mu_k} A^{\mu_1 \cdots \mu_k} e_{\mu_1} \wedge \cdots \wedge e_{\mu_k} \right],$$

from which we can calculate that ΛV has dimension 2^n.

3.1.4   *The Hodge star*

A pseudo inner product determines orthonormal bases for V, among which we can choose a specific one $\hat{e}_\mu$. The ordering of the $\hat{e}_\mu$ determines a choice of orientation. This orientation uniquely determines an orthonormal basis (i.e. a unit "length" vector) for the one-dimensional vector space $\Lambda^n V$, namely the **unit n-vector** (AKA orientation n-vector, volume element)

$$\Omega \equiv \hat{e}_1 \wedge \cdots \wedge \hat{e}_n.$$

$\triangle$ Many symbols are used in the literature for the unit n-vector and related quantities, including ε, i, I, and ω; to avoid confusion with the other common uses of these symbols, in this book we use the (non-standard) symbol Ω.

Since $\Lambda^n V$ is one-dimensional, every element of $\Lambda^n V$ is a real multiple of Ω. Thus Ω sets up a bijection (dependent upon the inner product and choice of orientation) between $\Lambda^n V$ and $\Lambda^0 V = \mathbb{R}$. In general, $\Lambda^k V$ and $\Lambda^{n-k} V$ are vector spaces of equal dimension, and thus we can also set up a bijection between them.

The **Hodge star operator** (AKA Hodge dual) is defined to be the linear map $*\colon \Lambda^k V \to \Lambda^{n-k} V$ that acts on $A \in \Lambda^k V$ such that for any $B \in \Lambda^{n-k} V$, we have

$$A \wedge B = \langle *A, B \rangle \, \Omega.$$

An equivalent requirement is that $\langle C \wedge *A, \Omega \rangle = \langle C, A \rangle$ for any $C \in \Lambda^k V$. In particular, we immediately obtain

$$A \wedge *A = \langle *A, *A \rangle \, \Omega = (-1)^s \, \langle A, A \rangle \, \Omega.$$

☼ These relations allow one to think of the Hodge star $*$ as an operator that roughly "swaps the exterior and inner products," or alternatively that yields the "orthogonal complement with the same magnitude."

The Hodge star operator is dependent upon a choice of inner product and orientation, but beyond that is independent of any particular basis. In particular, for any orthonormal basis $\hat{e}_\mu$ oriented with Ω, we can take $A \equiv \hat{e}_1 \wedge \cdots \wedge \hat{e}_k$ and $B \equiv \hat{e}_{k+1} \wedge \cdots \wedge \hat{e}_n$, in which case $*A = \langle B, B \rangle B$, i.e. $*A$ is constructed from an orthonormal basis for the orthogonal complement of A; in fact, this relation can be used as an equivalent definition of the Hodge star.

Below we list some easily derived facts about the Hodge star operator, where V is n-dimensional with unit n-vector Ω and a pseudo inner product of signature (r, s):

- $*\Omega = 1 \Rightarrow (*A) \, \Omega = A$ if $A \in \Lambda^n V$
- $*1 = (-1)^s \, \Omega \Rightarrow \langle \Omega, \Omega \rangle = (-1)^s$
- $**A = (-1)^{k(n-k)+s} A = (-1)^{k(n-1)+s} A$, where $A \in \Lambda^k V$
- $A \wedge *B = B \wedge *A = (-1)^s \, \langle A, B \rangle \, \Omega$ if $A, B \in \Lambda^k V$
- $\langle A, B \rangle = (-1)^s * (A \wedge *B)$ if $A, B \in \Lambda^k V$

△ Some authors instead define the Hodge star by the relation $A \wedge *B = \langle A, B \rangle \, \Omega$, which differs by a sign in some cases from the more common definition we use; in particular, with this definition $*\Omega = (-1)^s$ and $*1 = \Omega$.

Note that $*A$ is not a basis-independent object, since it reverses sign upon changing the chosen orientation. Such an object is prefixed by the word **pseudo-**, e.g. $*v$ is called a **pseudo-vector** (AKA axial vector, in which case v is called a polar vector) and Ω itself is a **pseudo-scalar**.

$\triangle$ The use of "pseudo" to indicate a quantity that reverses sign upon a change of orientation should not be confused with the use of "pseudo" to indicate an inner product that is not positive-definite. There are also other uses of "pseudo" in use. In particular, in general relativity the term "pseudo-tensor" is sometimes used, where neither of the above meanings are implied; instead this signifies that the tensor (to be defined in 3.2.2) is not in fact a tensor.

3.1.5 Graded algebras

The exterior algebra is an example of a **graded algebra**, which means that it has a decomposition, or **gradation** (AKA grading), into a direct sum of vector subspaces $\bigoplus V_g$ where each V_g corresponds to a **weight** (AKA degree), an element g of a monoid G (e.g. $\mathbb{N}$ under $+$) such that $V_g V_h = V_{g+h}$. The tensor algebra is a $\mathbb{N}$-graded algebra, since $T^j V \otimes T^k V = T^{j+k} V$, as is the exterior algebra of $\mathbb{R}^n$ (although V_j vanishes for $j > n$). The property $A \wedge B = (-1)^{jk} B \wedge A$ is then called **graded commutativity** (AKA graded anti-commutativity), whose definition can be generalized to other monoids.

A graded Lie algebra also obeys graded versions of the Jacobi identity and anti-commutativity. If we indicate the weight of v by $|v|$, the graded Lie bracket becomes $[u, v] = (-1)^{|u||v|+1} [v, u]$, and the graded Jacobi identity is $(-1)^{|u||w|} [[u, v] w] + (-1)^{|v||u|} [[v, w] u] + (-1)^{|w||v|} [[w, u], v] = 0$. A **Lie superalgebra** (AKA super Lie algebra) is a $\mathbb{Z}_2$-graded Lie algebra $V_0 \oplus V_1$ that is used to describe supersymmetry in physics.

3.1.6 Clifford algebras

Given an n-dimensional real vector space V with a pseudo inner product, we can generate another associative algebra called the **Clifford algebra** (AKA geometric algebra). As we will see, this algebra subsumes both the exterior algebra and the Hodge star.

The Clifford algebra generated by V is defined to be the tensor algebra modulo the identification $vv \equiv \langle v, v \rangle$, where juxtaposition denotes the vector multiplication operation on the algebra, called **Clifford multiplication** (AKA geometric multiplication). As a vector space, the Clifford algebra is isomorphic to the exterior algebra; in fact, the exterior product can be defined in terms of the Clifford product, leading immediately to simple relationships between these and the inner product:

- $\langle v, w \rangle = (vw + wv)/2$
- $v \wedge w \equiv (vw - wv)/2$
- $vw = \langle v, w \rangle + v \wedge w$

Each vector v can be written in terms of an orthonormal basis $\hat{e}_\mu$ of V, and for any pair of orthogonal vectors we have $\hat{e}_1\hat{e}_2 = \hat{e}_1 \wedge \hat{e}_2 = -\hat{e}_2\hat{e}_1$. The exterior product then naturally extends to any number of vectors by taking their completely anti-symmetrized sum under Clifford multiplication:

$$v_1 \wedge v_2 \wedge \cdots \wedge v_k \equiv \frac{1}{k!} \sum_\pi \text{sign}\,(\pi)\, v_{\pi(1)} v_{\pi(2)} \cdots v_{\pi(k)}$$

This completes the definition of the exterior product in terms of Clifford multiplication, carrying over all its properties from the exterior algebra. In particular, the Clifford algebra can be given the same basis as the orthonormal basis in the exterior algebra

$$\bigcup_{k=0}^{n} \{\hat{e}_{\mu_1} \wedge \cdots \wedge \hat{e}_{\mu_k}\}_{1 \leq \mu_1 < \cdots < \mu_k \leq n}$$

where we take $\{1\}$ for $k = 0$.

Given a choice of orientation with corresponding unit n-vector Ω, the Hodge star of an element $A \in \Lambda^k V$ may be written in terms of the Clifford product as

$$*A = (-1)^{\frac{k(k-1)}{2}+s} A\Omega$$

where the pseudo inner product has signature (r, s). It is helpful to see how the Hodge star works out in terms of the Clifford product in common signatures.

Table 3.1.1 The Hodge dual in terms of Clifford products in common signatures.

	$(2,0)$	$(3,0)$	$(3,1)$	$(1,3)$
$*v$	$-\Omega v = v\Omega$	$\Omega v = v\Omega$	$\Omega v = -v\Omega$	$\Omega v = -v\Omega$
$*B$	$-\Omega B = -B\Omega$	$-\Omega B = -B\Omega$	$\Omega B = B\Omega$	$\Omega B = B\Omega$
$*T$		$-\Omega T = -T\Omega$	$-\Omega T = T\Omega$	$-\Omega T = T\Omega$

Notes: here we have $v \in V$, $B \in \Lambda^2 V$, and $T \in \Lambda^3 V$.

In particular, the three dimensional cross product is seen to produce the pseudo-vector $v \times w = *(v \wedge w) = -\Omega (v \wedge w) = -(v \wedge w) \Omega$.

✿ Considering $ae_1(be_1 + ce_2) = ab + ac(e_1 \wedge e_2)$, we can think of Clifford multiplication as an operation that "scalar multiplies parallel components and exterior multiplies orthogonal ones." In particular, the product $A\Omega$ will "turn all basis vectors in A into scalars," yielding a form of orthogonal complement, as is the Hodge star.

3.1.7 *Geometric algebra*

The term "geometric algebra" usually refers to a relatively recent resurgence of interest in Clifford algebras, with an emphasis on geometric interpretations and motivations, and a variety of newly defined operations on the algebra. Here we will provide a brief synopsis of some of these ideas, with an eye towards potential usage in physical models.

△ It should be noted that definitions, notation, and terminology in geometric algebra vary quite significantly from author to author.

A general element of the Clifford algebra is called a **multivector** (AKA Clifford number), while a general element of $\Lambda^k V$ is called a **k-vector** (AKA homogeneous multivector) and is said to be of **grade** k (note that this is its weight as an element of the algebra graded over the exterior product, but not over the Clifford product).

An element that can be written as the exterior product of k vectors (or equivalently, as the Clifford product of k orthogonal vectors) is called a **k-blade**. So for example, $e_1 \wedge (e_2 + e_3)$ is a 2-blade, while $(e_1 \wedge e_2) + (e_3 \wedge e_4)$ is a 2-vector and $e_1 + (e_2 \wedge e_3)$ is a multivector. Note that if V has dimension 3 or less, every k-vector is a k-blade; in higher dimensions, they are in general sums of k-blades. 2-vectors are sometimes referred to as **bivectors**, and 3-vectors as **trivectors**.

The k-vector part of a general Clifford algebra element A is denoted $\langle A \rangle_k$ (the scalar part $\langle A \rangle_0$ is often written $\langle A \rangle$), and is a sum of k-blades. Thus any multivector can be decomposed into a sum of k-vectors: $A = \sum \langle A \rangle_k$. The unit n-vector Ω associated with a choice of orientation on V is sometimes called the **pseudo-scalar**.

Various operations can be defined on the entire Clifford algebra by defining them for k-blades and then using linearity. Below, we consider an a-blade A and a b-blade B.

- Dot product (AKA inner product): $A \bullet B \equiv \langle AB \rangle_{|a-b|}$ (the lowest grade part of the Clifford product AB)
- Exterior product (AKA outer product): $A \wedge B \equiv \langle AB \rangle_{(a+b)}$ (the highest grade part of AB)
- Reversion: the **reverse** $\widetilde{A}$ of $A = v_1 \wedge v_2 \wedge \cdots \wedge v_a$ reverses the order of its components $v_i \Rightarrow \widetilde{A} = (-1)^{a(a-1)/2} A$, $\widetilde{AB} = \widetilde{B}\widetilde{A}$

The definition of exterior product here can easily be shown to be equivalent to the usual one given in the exterior algebra, and thus shares the same symbol. In contrast, $A \bullet B$ is similar but not identical to the inner product $\langle A, B \rangle \equiv \det(\langle v_i, w_j \rangle)$ defined on the exterior algebra: $A \bullet B$ does not vanish for two elements of different grade, and for two k-blades one obtains the result $A \bullet B = \left\langle \widetilde{A}, B \right\rangle = (-1)^{k(k-1)/2} \langle A, B \rangle$. $\widetilde{A}$ is sometimes denoted $A^\dagger$, since under any representation of the Clifford algebra generated by hermitian matrices as vectors, the reverse corresponds to the hermitian conjugate.

Various relations then follow from these definitions. Below, we consider an a-blade A and a b-blade B, where V is n-dimensional and the pseudo inner product has signature (r, s).

- $v \bullet A = (vA - (-1)^a Av)/2$
- $v \wedge A = (vA + (-1)^a Av)/2 \Rightarrow vA = v \bullet A + v \wedge A$
- In particular, $v_0 \bullet (v_1 \wedge v_2) = (v_0 \bullet v_1) v_2 - (v_0 \bullet v_2) v_1$, and
- $v_0 \bullet (v_1 \wedge v_2 \wedge v_3) =$
  $(v_0 \bullet v_1)(v_2 \wedge v_3) - (v_0 \bullet v_2)(v_1 \wedge v_3) + (v_0 \bullet v_3)(v_1 \wedge v_2)$
- $\Omega \bullet A = \Omega A = (-1)^{a(n-1)} A\Omega = (-1)^{a(n-1)} A \bullet \Omega$
- $\Omega^2 \equiv \Omega\Omega = (-1)^{n(n-1)/2+s} \Rightarrow \widetilde{\Omega}\Omega = (-1)^s \Rightarrow *A = \left(\widetilde{\Omega}\Omega\right) \widetilde{A}\Omega$
- In particular, $\Omega^2 = -1$ for the signatures $(2,0), (3,0), (3,1)$ and $(1,3)$
- $\langle AB \rangle_0 = \langle BA \rangle_0 \Rightarrow \langle AB \cdots C \rangle_0 = \langle B \cdots CA \rangle_0$
- If B is a bivector, the commutator $[A, B] = AB - BA$ has the same grade as $A \Rightarrow$ the commutator of two bivectors is another bivector
- If B is a bivector, $BA = B \bullet A + [B, A]/2 + B \wedge A$

$\triangle$ It should be noted that some authors in geometric algebra define a "dual" that differs by a (grade- and signature-dependent) sign from the usual Hodge dual used in most texts. For example, $A\Omega$ is sometimes defined as the "dual" of A, in which case the "swapping of the inner and exterior products" property can be generalized to the form $A \bullet (B\Omega) = (A \wedge B)\,\Omega$, valid for all $a + b \leq n$.

3.2 Tensor algebras on the dual space

Given a finite-dimensional vector space V, the **dual space** V^* is defined to be the set of linear mappings from V to the scalars (AKA the linear functionals on V). The elements of V^* can be added together and multiplied by scalars, so V^* is also a vector space, with the same dimension as V.

$\triangle$ Note that in general, the word "dual" is used for many concepts in mathematics; in particular, the dual space has no relation to the Hodge dual.

3.2.1 *The structure of the dual space*

An element $\varphi\colon V \to \mathbb{R}$ of V^* is called a **1-form**. Given a pseudo inner product on V, we can construct an isomorphism between V and V^* defined by $v \mapsto \langle v,\ \rangle$, i.e. $v \in V$ is mapped to the element of V^* which maps any vector $w \in V$ to $\langle v, w \rangle$. This isomorphism then induces a corresponding pseudo inner product on V^* defined by $\langle \langle v,\ \rangle, \langle w,\ \rangle \rangle \equiv \langle v, w \rangle$.

An equivalent way to set up this isomorphism is to choose a basis e_μ of V, and then form the **dual basis** β^λ of V^*, defined to satisfy $\beta^\lambda(e_\mu) = \delta^\lambda{}_\mu$. The isomorphism between V and V^* is then defined by the correspondence $v = v^\mu e_\mu \mapsto (\eta_{\lambda\mu}v^\mu)\beta^\lambda \equiv v_\lambda \beta^\lambda$, corresponding to the isomorphism induced by the pseudo inner product on V that makes e_μ orthonormal. Note that if $\langle e_\mu, e_\mu \rangle = -1$ then $e_\mu \mapsto -\beta^\mu$. This isomorphism and its inverse (usually in the context of Riemannian manifolds) are called the **musical**

isomorphisms, where if $v = v^\mu e_\mu$ and $\varphi = \varphi_\mu \beta^\mu$ we write

$$v^\flat \equiv \langle v, \ \rangle$$
$$= \left(\eta_{\mu\lambda} v^\lambda \right) \beta^\mu$$
$$\equiv v_\mu \beta^\mu$$
$$\varphi^\sharp \equiv \langle \varphi, \ \rangle$$
$$= \left(\eta^{\mu\lambda} \varphi_\lambda \right) e_\mu$$
$$\equiv \varphi^\mu e_\mu$$

and call the $v^\flat$ the **flat** of the vector v and $\varphi^\sharp$ the **sharp** of the 1-form φ.

☼ A 1-form acting on a vector can thus be viewed as yielding a projection. Specifically, $\varphi(v)/\left\|\varphi^\sharp\right\|$ is the length of the projection of v onto the ray defined by $\varphi^\sharp$.

It is important to note that there is no **canonical isomorphism** between V and V^*, i.e. we cannot uniquely associate a 1-form with a given vector without introducing extra structure, namely an inner product or a preferred basis. Either structure will do: a choice of basis is equivalent to the definition of the unique inner product on V that makes this basis orthonormal, which then induces the same isomorphism as that induced by the dual basis.

In contrast, a canonical isomorphism $V \cong V^{**}$ can be made via the association $v \in V \leftrightarrow \xi \in V^{**}$ with $\xi \colon V^* \to \mathbb{R}$ defined by $\xi\left(\varphi\right) \equiv \varphi\left(v\right)$. Thus V and V^{**} can be completely identified (for a finite-dimensional vector space), and we can view V as the dual of V^*, with vectors regarded as linear mappings on 1-forms.

Note that since $\beta^\lambda(e_\mu) = \delta^\lambda{}_\mu$ and $\langle e_\mu, e_\lambda \rangle = \eta^{\mu\lambda}$ we have

$$\varphi(v) = \varphi_\lambda \beta^\lambda(v^\mu e_\mu)$$
$$= \varphi_\mu v^\mu$$
$$= \eta_{\mu\lambda} \varphi^\lambda v^\mu \langle e_\mu, e_\lambda \rangle$$
$$= \langle \varphi^\sharp, v \rangle.$$

Vector components are often viewed as a column vector, which means that 1-forms act on vector components as row vectors (which then are acted on by matrices from the right). Under a change of basis we then have the following relationships:

Table 3.2.1 Transformations under a change of basis.

	Index notation	Matrix notation
Basis	$e'_\mu = A^\lambda{}_\mu e_\lambda$	$e' = eA$
Dual basis	$\beta'^\mu = (A^{-1})^\mu{}_\lambda \beta^\lambda$	$\beta' = A^{-1}\beta$
Vector components	$v'^\mu = (A^{-1})^\mu{}_\lambda v^\lambda$	$v' = A^{-1}v$
1-form components	$\varphi'_\mu = A^\lambda{}_\mu \varphi_\lambda$	$\varphi' = \varphi A$

> Notes: A 1-form will sometimes be viewed as a column vector, i.e. as the transpose of the row vector (which is the sharp of the 1-form under a Riemannian signature). Then we have $(\varphi')^T = (\varphi A)^T = A^T \varphi^T$.

3.2.2 Tensors

A **tensor** of **type** (AKA valence) (m,n) is defined to be an element of the tensor space

$$V_{m,n} \equiv (V \otimes \cdots (m\text{ times}) \cdots \otimes V) \otimes (V^* \otimes \cdots (n\text{ times}) \cdots \otimes V^*).$$

A **pure tensor** (AKA simple or decomposable tensor) of type (m,n) is one that can be written as the tensor product of m vectors and n 1-forms; thus a general tensor is a sum of pure tensors. The integer $(m+n)$ is called the **order** (AKA degree, rank) of the tensor, while the tensor **dimension** is that of V. Vectors and 1-forms are then tensors of type $(1,0)$ and $(0,1)$. The **rank** (sometimes used to refer to the order) of a tensor is the minimum number of pure tensors required to express it as a sum. In "tensor language" vectors $v \in V$ are called **contravariant vectors** and 1-forms $\varphi \in V^*$ are called **covariant vectors** (AKA covectors). Scalars can be considered tensors of type $(0,0)$.

3.2.3 Tensors as multilinear mappings

There is an obvious multiplication of two 1-forms: the scalar multiplication of their values. The resulting object $\varphi\psi\colon V \times V \to \mathbb{R}$ is a nondegenerate bilinear form on V. Viewed as an "outer product" on V^*, multiplication is trivially seen to be a bilinear operation, i.e. $a(\varphi + \psi)\xi = a\varphi\xi + a\psi\xi$. Thus the product of two 1-forms is isomorphic to their tensor product.

We can extend this to any tensor by viewing vectors as linear mappings on 1-forms, and forming the isomorphism $\bigotimes \varphi_i \mapsto \prod \varphi_i$. Note that this isomorphism is not unique, since for example any real multiple of the product would yield a multilinear form as well. However it is canonical, since the choice does not impose any additional structure, and is also consistent with considering scalars as tensors of type $(0,0)$.

We can thus consider tensors to be multilinear mappings on V^* and V. In fact, we can view a tensor of type (m,n) as a mapping from $i < m$ 1-forms and $j < n$ vectors to the remaining $(m-i)$ vectors and $(n-j)$ 1-forms.

$$[\ w \otimes \psi \otimes \xi\ ]\ (\ \varphi,\ v,\ \ )\ =\ \varphi(w)\cdot\psi(v)\ [\ \xi\ ]$$

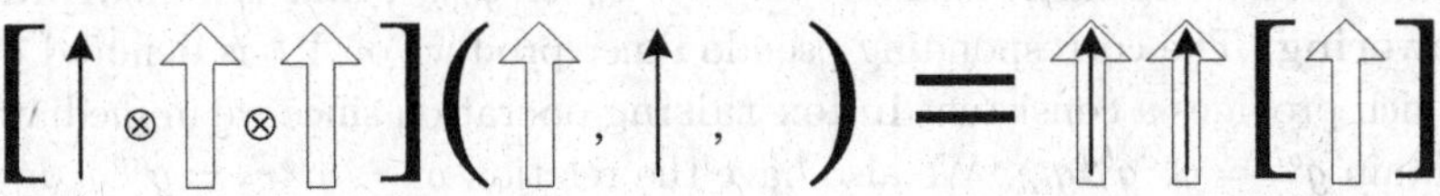

$$\left[\uparrow\cdot\Uparrow\otimes\Uparrow\right]\left(\Uparrow,\uparrow,\ \right)=\Uparrow\Uparrow\left[\Uparrow\right]$$

$$T\qquad(\ \varphi,\ v,\ \ )\qquad=\qquad \zeta$$

$$T^{u}_{\ bc}\qquad \varphi_a\ v^b\qquad\qquad =\qquad \zeta_c$$

Figure 3.2.1 Different ways of depicting a pure tensor of type $(1,2)$. The first line explicitly shows the tensor as a mapping from a 1-form φ and a vector v to a 1-form ξ. The second line visualizes vectors as arrows, and 1-forms as receptacles that when matched to an arrow yield a scalar. The third line combines the constituent vectors and 1-forms of the tensor into a single symbol T while merging the scalars into ξ to define ζ, and the last line adds indices (covered in the next section).

A general tensor is a sum of pure tensors, so for example a tensor of the form $(u\otimes\varphi)+(v\otimes\psi)$ can be viewed as a linear mapping that takes ξ and w to the scalar $\xi(u)\cdot\varphi(w)+\xi(v)\cdot\psi(w)$. Since the roles of mappings and arguments can be reversed, we can simplify things further by viewing the arguments of a tensor as another tensor: $(u\otimes\varphi)(\xi\otimes w)\equiv(u\otimes\varphi)(\xi,w)=(\xi\otimes w)(u,\varphi)=\xi(u)\cdot\varphi(w)$.

3.2.4 Abstract index notation

Abstract index notation uses an upper Latin index to represent each contravariant vector component of a tensor, and a lower index to represent each covariant vector (1-form) component. We can see from the preceding figure that this notation is quite compact and clearly indicates the type of each tensor while re-using letters to indicate what "slots" are to be used in the mapping.

The tensor product of two tensors $S^a_{\ b}\otimes T^c_{\ d}$ is simply denoted $S^a_{\ b}T^c_{\ d}$, and in this form the operation is sometimes called the **tensor direct product**. We may also consider a **contraction** $T^{ab}_{bc} = T^{a}_{c}$, where two of the components of a tensor operate on each other to create a new tensor with a reduced number of indices. For example, if $T^{ab}_{c} = v^a \otimes w^b \otimes \varphi_c$, then $T^{ab}_{b} = \varphi(w) \cdot v^a$.

A (pseudo) inner product on V is a symmetric bilinear mapping, and thus corresponds to a symmetric tensor g_{ab} called the **(pseudo) metric tensor**. The isomorphism $v \in V \mapsto v^b \in V^*$ induced by this pseudo inner product is then defined by $v^a \mapsto v_a \equiv g_{ab}v^b$, and is called **index lowering**. The corresponding pseudo inner product on V^* is denoted g^{ab}, which provides a consistent **index raising** operation since we immediately obtain $g^{ab} = g^{ac}g^{bd}g_{cd}$. We also have the relation $v^a = g^{ab}v_b = g^{ab}g_{bc}v^c \Rightarrow g^{ab}g_{bc} = g^a_{c} = \delta^a_{c}$, the identity mapping. The inner product of two tensors of the same type is then the contraction of their tensor direct product after index lowering/raising, e.g. $\langle T^{ab}, S^{cd} \rangle = T^{ab}S_{ab} = T^{ab}g_{ac}g_{bd}S^{cd}$.

$\triangle$ It is important to remember that if v is a vector, the operation $v_a v^a$ implies index lowering, which requires an inner product. In contrast, if φ is a 1-form, the operation $\varphi_a v^a$ is always valid regardless of the presence of an inner product.

A symmetric or anti-symmetric tensor can be formed from a general tensor by adding or subtracting versions with permuted indices. For example, the combination $(T_{ab} + T_{ba})/2$ is the **symmetrized tensor** of T, i.e. exchanging any two indices leaves it invariant. The **anti-symmetrized tensor** $(T_{ab} - T_{ba})/2$ changes sign upon the exchange of any two indices, and yields the original tensor T_{ab} when added to the symmetrized tensor. The following notation is common for tensors with n indices, with the sums over all permutations of indices:

$$\text{Symmetrization: } T_{(a_1 \dots a_n)} \equiv \frac{1}{n!} \sum_{\pi} T_{a_{\pi(1)} \dots a_{\pi(n)}}$$

$$\text{Anti-symmetrization: } T_{[a_1 \dots a_n]} \equiv \frac{1}{n!} \sum_{\pi} \text{sign}\,(\pi)\, T_{a_{\pi(1)} \dots a_{\pi(n)}}$$

This operation can be performed on any subset of indices in a tensor, as long as they are all covariant or all contravariant. Skipping indices is denoted with vertical bars, as in $T_{(a|b|c)} = (T_{abc} + T_{cba})/2$; however, note that vertical bars alone are sometimes used to denote a sum of ordered permutations, as in $T_{|abc|} = T_{abc} + T_{bca} + T_{cab}$.

3.2.5   Tensors as multi-dimensional arrays

In a given basis, a pure tensor of type (m, n) can be written using **component notation** in the form

$$v^1 \otimes \cdots \otimes v^m \otimes \varphi_1 \otimes \cdots \otimes \varphi_n \equiv T^{\mu_1 \ldots \mu_m}{}_{\nu_1 \ldots \nu_n} e_{\mu_1} \otimes \cdots \otimes e_{\mu_m} \otimes \beta^{\nu_1} \otimes \cdots \otimes \beta^{\nu_n},$$

where the Einstein summation convention is used in the second expression. Note that the collection of terms into T is only possible due to the defining property of the tensor product being linear over addition. The tensor product between basis elements is often dropped in such expressions.

A general tensor is a sum of such pure tensor terms, so that any tensor T can be represented by a $(m + n)$-dimensional array of scalars. For example, any tensor of order 2 is a matrix, and as a linear mapping operates on vectors and forms via ordinary matrix multiplication if they are all expressed in terms of components in the same basis. Basis-independent quantities from linear algebra such as the trace and determinant are then well-defined on such tensors.

> $\triangle$ A potentially confusing aspect of component notation is the basis vectors e_μ, which are not components of a 1-form but rather vectors, with μ a label, not an index. Similarly, the basis 1-forms β^ν should not be confused with components of a vector.

The Latin letters of abstract index notation (e.g. $T^{ab}{}_{cd}$) can thus be viewed as placeholders for what would be indices in a particular basis, while the Greek letters of component notation represent an actual array of scalars that depend on a specific basis. The reason for the different notations is to clearly distinguish tensor identities, true in any basis, from equations true only in a specific basis.

> $\triangle$ In general relativity both abstract and index notation are abused to represent objects that are non-tensorial. We will see this in Chapter 9.

△ Note that if abstract index notation is not being used, Latin and Greek indices are often used to make other distinctions, a common one being between indices ranging over three space indices and indices ranging over four spacetime indices.

△ Note that "rank" and "dimension" are overloaded terms across these constructs: "rank" is sometimes used to refer to the order of the tensor, which is the dimensionality of the corresponding multi-dimensional array; the dimension of a tensor is that of the underlying vector space, and so is the length of a side of the corresponding array (also sometimes called the dimension of the array). However, the rank of a order 2 tensor coincides with the rank of the corresponding matrix.

3.3 Exterior forms

Elements of the exterior algebra of the dual space can be viewed in several ways: as mappings on V, as tensors, or as arrays.

△ As we will see, these correspondences are not unique, and in the literature they are often glossed over or conventions are implicitly assumed.

3.3.1 *Exterior forms as multilinear mappings*

An **exterior form** (AKA k-form, alternating form) is defined to be an element of $\Lambda^k V^*$. Just as we formed the isomorphism $\otimes \varphi_i \mapsto \Pi \varphi_i$ to view tensors as multilinear mappings on V, we can view k-forms as alternating multilinear mappings on V. Restricting attention to the exterior product of k 1-forms $\bigwedge \varphi_i$, we define the isomorphism

$$
\bigwedge_{i=1}^{k} \varphi_i \mapsto \sum_{\pi} \operatorname{sign}(\pi) \prod_{i=1}^{k} \varphi_{\pi(i)}
$$

$$
= \sum_{i_1, i_2, \ldots, i_k} \varepsilon^{i_1 i_2 \ldots i_k} \varphi_{i_1} \varphi_{i_2} \cdots \varphi_{i_k} = \varepsilon^I \varphi_I,
$$

where we recall the combinatorial notations from Section 3.1.3.

> ✫ The above isomorphism extends the interpretation of forms acting on vectors as yielding a projection. Specifically, if the parallelepiped $\varphi^{\sharp} = \bigwedge \varphi_i^{\sharp}$ has volume V, then $\varphi(v_1, \ldots v_k)/V$ is the volume of the projection of the parallelepiped $v = \bigwedge v_i$ onto $\varphi^{\sharp}$.

Extending this to arbitrary forms $\varphi \in \Lambda^j V^*$ and $\psi \in \Lambda^k V^*$, we have

$$(\varphi \wedge \psi)(v_1, \ldots, v_{j+k})$$

$$\mapsto \frac{1}{j!k!} \sum_{\pi} \text{sign}(\pi)\, \varphi\left(v_{\pi(1)}, \ldots, v_{\pi(j)}\right) \psi\left(v_{\pi(j+1)}, \ldots, v_{\pi(j+k)}\right).$$

Just as with tensors, this isomorphism is canonical but not unique; but in the case of exterior forms, other isomorphisms are in common use. The main alternative isomorphism inserts a term $1/k!$ in the first relation above, which results in $1/j!k!$ being replaced by $1/(j+k)!$ in the second. Note that this alternative is inconsistent with the interpretation of exterior products as parallelepipeds.

> △ It is important to understand which convention a given author is using. The first convention above is common in physics, and we will adhere to it in this book.

3.3.2 Exterior forms as completely anti-symmetric tensors

An immediate result of this view of forms as multilinear mappings is that we can also view forms as completely anti-symmetric tensors under the identification of $\prod \varphi_i$ with $\bigotimes \varphi_i$. For example, for a 2-form we have the equivalent expressions $(\varphi \wedge \psi)(v, w) \leftrightarrow (\varphi \otimes \psi - \psi \otimes \varphi)(v, w) \leftrightarrow \varphi(v)\psi(w) - \psi(v)\varphi(w)$.

Note however that this identification does not lead to equality of the inner products defined on tensors and exterior forms; instead for two k-forms we have

$$\left\langle \bigwedge \varphi_i, \bigwedge \psi_j \right\rangle_{\text{form}} = \det\left((\langle \varphi_i, \psi_j \rangle)\right),$$

while as tensors we have

$$\left\langle \bigwedge \varphi_i, \bigwedge \psi_j \right\rangle_{\text{tensor}} = \left\langle \varepsilon^I \varphi_I, \varepsilon^J \varphi_J \right\rangle_{\text{tensor}} = k! \, (k-1)! \det\left(\langle \varphi_i, \psi_j \rangle\right).$$

Fortunately, the tensor inner product is almost always expressed explicitly in terms of index contractions, so we will continue to use the $\langle \, , \, \rangle$ notation for the inner product of k-forms.

3.3.3 Exterior forms as anti-symmetric arrays

In terms of a basis β^μ of V^*, we can write a k-form φ as

$$\varphi = \frac{1}{k!} \sum_{\mu_1,\ldots,\mu_k} \varphi_{\mu_1\ldots\mu_k} \beta^{\mu_1} \wedge \cdots \wedge \beta^{\mu_k}.$$

$\triangle$ The above way of writing the components is not unique, and others are in common use, the main alternative omitting the factorial.

The advantage of the expression above is that, with our isomorphism convention, the component array can be identified with the anti-symmetric covariant tensor component array in the same basis:

$$\varphi = \frac{1}{k!} \varphi_{\mu_1\ldots\mu_k} \sum_\pi \text{sign}\,(\pi) \prod_i \beta^{\pi(i)} = \varphi_{\mu_1\ldots\mu_k} \beta^{\mu_1\ldots\mu_k}$$

Here we have dropped the summation sign in favor of the Einstein summation convention, and the last equality follows from the anti-symmetry of the component array.

3.3.4 The Clifford algebra of the dual space

An immediately apparent notational issue with the Clifford algebra of V^* is that juxtaposition is used to denote both Clifford multiplication and the multiplication of the scalar values of forms. However, they can be distinguished if the arguments are explicitly noted, e.g. the Clifford product of 1-forms is written $\varphi\psi\,(v,w)$ versus the scalar product of values $\varphi\,(v)\,\psi\,(w)$.

We can use the view of exterior forms as mappings to view the Clifford product of 1-forms as $\varphi\psi\,(v,w) = \langle \varphi, \psi \rangle + (\varphi \wedge \psi)\,(v,w) \mapsto \langle \varphi, \psi \rangle + \varphi\,(v)\,\psi\,(w) - \psi\,(v)\,\varphi\,(w)$. Note that in this view the Clifford product is not a multilinear mapping on V, since there is a leading constant; it is an affine mapping (defined in Section 7.4.5). The Clifford multiplication of higher-grade j- and k-vectors yields a constant plus multivectors of mixed grade, which in this view are then a sum of multilinear mappings on subsets of the $(j + k)$ vector arguments.

3.3.5 Algebra-valued exterior forms

We can extend the view of exterior forms as real-valued linear mappings to define **algebra-valued forms**. These follow the same construction as in Section 3.3.1 above, starting from an algebra-valued 1-form $\check{\Theta} \colon V \to \mathfrak{a}$, so that general forms are alternating multilinear maps from k vectors to a real algebra $\mathfrak{a}$ whose vector multiplication takes the place of multiplication in $\mathbb{R}$. Since this vector multiplication may not be commutative, we need to be more careful in terms of ordering in the isomorphism to ensure antisymmetry, i.e. for two algebra-valued 1-forms we define

$$(\check{\Theta} \wedge \check{\Psi})(v, w) \equiv \check{\Theta}(v)\check{\Psi}(w) - \check{\Theta}(w)\check{\Psi}(v).$$

An algebra-valued form whose values are defined by matrices is a **matrix-valued form**. Exterior forms that take values in a matrix group can also be considered as matrix-valued forms, but it must be understood that under addition the values may no longer be in the group. One can also form the exterior product between a matrix-valued form and a **vector-valued form**. To reduce confusion when dealing with algebra- and vector-valued forms, we will indicate them with (non-standard) decorations, for example in the case of a matrix-valued 1-form acting on a vector-valued 1-form,

$$(\check{\Theta} \wedge \vec{\varphi})(v, w) \equiv \check{\Theta}(v)\vec{\varphi}(w) - \check{\Theta}(w)\vec{\varphi}(v).$$

△ An additional distinction can be made between forms that take values which are concrete matrices and column vectors (and thus depend upon the basis of the underlying vector space), and forms that take values which are abstract linear transformations and abstract vectors (and thus are basis-independent). We will attempt to distinguish between these by referring to the specific matrix or abstract group, and by only using "vector-valued" when the value is an abstract vector.

A notational issue arises in the particular case of Lie algebra valued forms, where the related associative algebra in the relation $[\check{\Theta}, \check{\Psi}] = \check{\Theta}\check{\Psi} -$

$\check{\Psi}\check{\Theta}$ could also be in use. In this case multiplication of values could use
either the Lie commutator or that of the related associative algebra. We
will denote the exterior product using the Lie commutator by $\check{\Theta}[\wedge]\check{\Psi}$. Some
authors use $[\check{\Theta},\check{\Psi}]$ or $[\check{\Theta}\wedge\check{\Psi}]$, but both can be ambiguous, motivating us
to introduce our (non-standard) notation. The expression $\check{\Theta}\wedge\check{\Psi}$ is then
reserved for the exterior product using the underlying associative algebra
(e.g. that of matrix multiplication if the associative algebra is defined this
way). For two Lie algebra-valued 1-forms we then have

$$(\check{\Theta}[\wedge]\check{\Psi})(v,w) = [\check{\Theta}(v),\check{\Psi}(w)] - [\check{\Theta}(w),\check{\Psi}(v)]$$
$$= \check{\Theta}(v)\check{\Psi}(w) - \check{\Psi}(w)\check{\Theta}(v) - \check{\Theta}(w)\check{\Psi}(v) + \check{\Psi}(v)\check{\Theta}(w).$$

Note that $[\check{\Theta},\check{\Psi}](v,w) = \check{\Theta}(v)\check{\Psi}(w) - \check{\Psi}(v)\check{\Theta}(w)$ is a distinct construction,
as is $[\check{\Theta}(v),\check{\Psi}(w)] = \check{\Theta}(v)\check{\Psi}(w) - \check{\Psi}(w)\check{\Theta}(v)$; neither are in general anti-
symmetric and thus do not yield forms. Also note that e.g. for two 1-forms
$\check{\Theta}[\wedge]\check{\Psi} \neq \check{\Theta}\wedge\check{\Psi} - \check{\Psi}\wedge\check{\Theta}$, and $(\check{\Theta}[\wedge]\check{\Theta})(v,w) = 2[\check{\Theta}(v),\check{\Theta}(w)]$ does not in
general vanish, so $[\wedge]$ does not act like a Lie commutator in these respects.
Instead it forms a graded Lie algebra, so that for algebra-valued j- and
k-forms $\check{\Theta}$ and $\check{\Psi}$ we have the graded commutativity rule

$$\check{\Theta}[\wedge]\check{\Psi} = (-1)^{jk+1}\,\check{\Psi}[\wedge]\check{\Theta},$$

and with an algebra-valued m-form $\check{\Xi}$ we have the graded Jacobi identity

$$(-1)^{jm}(\check{\Theta}[\wedge]\check{\Psi})[\wedge]\check{\Xi} + (-1)^{kj}(\check{\Psi}[\wedge]\check{\Xi})[\wedge]\check{\Theta} + (-1)^{mk}(\check{\Xi}[\wedge]\check{\Theta})[\wedge]\check{\Psi} = 0.$$

Algebra-valued forms also introduce potentially ambiguous index notation. If a basis is chosen for the algebra $\mathfrak{a}$, the value of an algebra-valued
form may be expressed using component notation as Θ^{μ}; or if the algebra
is defined in terms of matrices, an element might be written $\Theta^{\alpha}{}_{\beta}$, an expression that has nothing to do with the basis of $\mathfrak{a}$. Then for example an
algebra-valued 1-form might be written $\Theta^{\mu}{}_{\gamma}$ or $\Theta^{\alpha}{}_{\beta\gamma}$.

$\triangle$ In considering algebra-valued forms expressed in index notation,
extra care must be taken to identify the type of form in question, and
to match each index with the aspect of the object it was meant to
represent.

3.3.6 Related constructions and facts

Exterior forms are usually simply called "forms," but as we saw in Section 2.2.3 this term is also used to describe general multilinear mappings from vector spaces to scalars. Here we revisit these mappings to give their equivalent definitions in "tensor language."

- Bilinear form: a covariant tensor of order 2
- Multilinear form: a covariant tensor of arbitrary order
- Quadratic form: a symmetric covariant tensor of order 2
- Algebraic form: a symmetric covariant tensor of arbitrary order
- Exterior form: an anti-symmetric covariant tensor of arbitrary order
- Pseudo inner product: a nondegenerate completely symmetric covariant tensor of order 2
- Symplectic form: a nondegenerate completely anti-symmetric covariant tensor of order 2

The operation of the exterior product by a fixed 1-form $(\varphi\wedge)$ can be viewed as a linear mapping from $\Lambda^k V^*$ to $\Lambda^{k+1} V^*$. We can form a mapping that goes in the opposite direction, the **interior product** of φ by a vector v, denoted $i_v\varphi$ (also denoted $v\lrcorner\varphi$). This operation fixes the first vector argument of a k-form, i.e. $(i_v\varphi)(w_2,\dots,w_k) \equiv \varphi(v,w_2,\dots,w_k)$. The interior product is thus a linear mapping from $\Lambda^k V^*$ to $\Lambda^{k-1} V^*$. We define $i_v f \equiv 0$ for a 0-form f and note that $i_v\Omega = *(v^\flat)$.

△ The relationship between the interior and exterior product is not that of "opposites," i.e. neither reverses the effect of the other. Instead, as we will see in Section 6.3.6, the interior product acts as a kind of derivative.

Chapter 4

Topological spaces

Using the algebraic tools we have developed, we can now move into geometry. Before launching into the main subject of this chapter, topology, we will examine the intuitive meanings of geometric objects in general, and the properties that define them.

4.1 Generalizing surfaces

In algebra we added arithmetic properties to sets cumulatively until we had built enough structure to arrive at the real numbers. We then went further by considering generalizations of vectors. In geometry, we build up to a surface in three-dimensional space, and then go further by considering generalizations of tangent vectors to a surface.

In this section we will give a preview of the basic ideas, with the implicit assumption that the reader already has some familiarity with objects and relations that will only be defined later.

Table 4.1.1 Generalizations of surfaces.

	Added structure	Resulting features
Topological space	Open sets	Connectedness, holes
Hausdorff space	Disjoint neighborhoods	Separation between points
Metric space	Metric	Distances between points
Topological manifold	Cartesian charts	Coordinates, dimension
Differentiable manifold	Differentiable structure	Tangent vectors, calculus
Riemannian manifold	Riemannian metric	Length, angles, volumes
Euclidean surface	Embedding in $\mathbb{R}^3$	Relation to a higher space

Notes: There are many more messy details here than with algebra. These will be covered in the following chapters.

4.1.1 *Spaces*

Let's start with a Euclidean surface and examine what happens as we discard various properties. A two-dimensional Riemannian surface only includes **intrinsic** information, i.e. information that is independent of any outside structure, and so may not have a unique embedding in $\mathbb{R}^3$. For example, a sheet of paper is flat, and remains intrinsically so even if it is rolled up; i.e. there is no measurement possible on the surface of the sheet that can reveal whether it is rolled up or not. A plain manifold includes no notion of length at all, so the sheet of paper can be arbitrarily stretched, as long as there are no kinks or singularities introduced. There are other objects that we will not consider here, but may have application to physics. For example, an **orbifold** is a manifold where certain kinds of singularities are allowed.

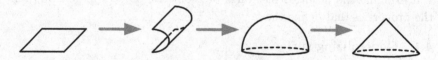

Figure 4.1.1 A manifold being rolled up, stretched, and kinked.

Short of a manifold structure we can define a series of increasingly more primitive spaces, but they begin to include examples that are hard to reconcile with our intuitive idea of a "geometric object." Our main purpose in studying these, in particular topological spaces, will be to determine what results only depend on "shape," i.e. what results only depend on topology, as opposed to an additional manifold structure.

☼ In this spirit, we will view topological spaces as manifolds with their Cartesian charts ignored.

4.1.2 *Generalizing dimension*

An additional aspect of surfaces that is generalized above is that of dimension. We are intuitively familiar with objects of dimension up to three: points, curves, surfaces, and volumes. As we build structure in geometry, we try to keep all our definitions applicable to any number of dimensions.

The Cartesian charts of a manifold determine an unambiguous dimensionality, since they are maps to $\mathbb{R}^n$ for a specific integer n. In contrast, short of a manifold structure there are several definitions of dimension, some of which result in non-integer values (called **fractal dimensions**). It is also important to note that properties that hold in easily visualized lower dimensional manifolds do not always remain valid in higher dimensions.

4.1.3 Generalizing tangent vectors

Beyond generalizing surfaces, we can generalize the properties of tangent vectors to a surface. The tangent vectors at each point of a surface lie in a two-dimensional space, a copy of $\mathbb{R}^2$ at each point. Accordingly, the tangent vectors of a general n-dimensional differentiable manifold are defined to lie in a copy of the vector space $\mathbb{R}^n$ at each point. As we will see in subsequent chapters, the space at each point is called the tangent space, and the tangent spaces at every point taken together are called the tangent bundle. A topological manifold lacks sufficient structure to define tangents.

We can generalize the idea of tangents further by considering a new arbitrary vector space to be associated with each point of a manifold, an "internal space" that has nothing to do with tangents. This is the basic idea behind gauge theories in physics, and is best described using the language of fiber bundles.

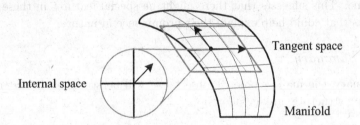

Figure 4.1.2 Generalizing tangents on a manifold.

4.1.4 Existence and uniqueness of additional structure

The interplay between successive geometrical structures is an active area of research with many unknowns. With each added structure we can ask several questions:

- Can the higher structure always be defined on any object?
- Can more than one non-equivalent structure be defined on a given object?
- Does the structure provide new ways to distinguish between objects?

The answer to the first question can be negative: for example, there exist topological spaces that admit no Cartesian charts, and topological manifolds that admit no differential structure (in dimension 4 or higher). On the other hand, all differentiable manifolds admit a Riemannian metric. The answer to the second question is usually positive: for example, one can define non-equivalent metrics on a given differential manifold, and even non-equivalent differentiable structures on a given topological manifold (e.g. Milnor's exotic 7-spheres). So we can conclude that in general, additional structure usually expands the possibilities, but also may eliminate spaces that are not "nice" enough.

The third question can also be positive, as seen in **Donaldson theory** and **Seiberg-Witten theory**, which use the structure of gauge theories in physics to create new ways of distinguishing 4-dimensional manifolds. These theories help show, among other things, that one can define non-equivalent differentiable structures on $\mathbb{R}^4$, a situation that is only true in four dimensions.

This brings us to a particularly interesting observation concerning geometrical structure, the fact that the answers to many questions often change or are most difficult to answer in dimension 3 and/or 4. These are of course the most common dimensionalities in physics, corresponding to space and spacetime. This suggests that there might be special features in these dimensions that could help explain their prominence in nature.

4.1.5   *Summary*

In summary, the main geometric objects we will be studying, in order of increasing structure, are as follows:

Topological space:
- Structure: Open sets
- Properties: Number of holes, connectedness
- Equivalency: Homeomorphism

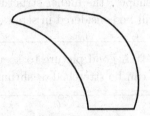

Topological manifold:
- Structure: Local cartesian charts
- Properties: Coordinates, dimension
- Equivalency: Homeomorphism

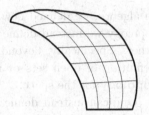

Differentiable manifold:
- Structure: Differentiable structure
- Properties: Tangent vectors, calculus
- Equivalency: Diffeomorphism

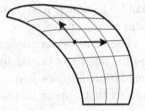

Riemannian manifold:
- Structure: Metric
- Properties: Lengths, angles, areas, volumes
- Equivalency: Isometry

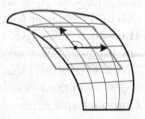

Figure 4.1.3 Geometric objects in order of increasing structure.

4.2 Generalizing shapes

We now consider the simplest structure in the preceding figure, the topological space. As is often done, if there is no risk of confusion we will shorten this to just "space." Spaces essentially only include the intuitive idea of "shape"; the higher structures and ideas present in manifolds and beyond will be considered in separate chapters.

☼ A good picture to keep in mind for a space is a rubber sheet that can be stretched or shrunk, but cannot be torn or glued.

4.2.1   Defining spaces

In algebra, we defined how to operate on two elements to get another; in topology we instead define in a rough sense how close points are to each other. Specifically, beyond being a set, a **topological space** includes a definition of open sets or neighborhoods. This is also called defining a **topology** for the space.

One can instead define a numeric distance between points, forming a **metric space**, which is automatically a topological space; but in physics it is generally more profitable to jump from topological spaces directly to manifolds, which have a natural distance function that can always make them metric spaces.

The more modern definition of a topology uses open sets, but equivalent results can be obtained by defining neighborhoods. We will not go into foundational topics here, but it is worthwhile to be reminded of a few notions that appear frequently concerning a space X:

- **Weaker** (AKA coarser) topology: one topology is weaker than another if it defines less open sets, i.e., every open set in the weaker topology is also in the stronger (AKA finer) one
- **Hausdorff** space (AKA T_2, satisfying the second "separation axiom"): any two points have disjoint neighborhoods, i.e. no two points are "right next to each other"; this is a common condition on spaces to avoid pathological exceptions
- **Compact** space: every cover has a finite sub-cover; we will not define these terms here, but this usually translates to closed and bounded
- **Connected** space: the only subsets that are both open and closed are X and the empty set; this implies the intuitive idea that X does not consist of several disjoint pieces
- **Continuous mapping** $f \colon X \to Y$: inverse images of open sets are open, i.e. if $B \subset Y$ is open, $f^{-1}(B) \subset X$ is open; this implies the intuitive properties of continuity

- **Path-connected** space: a more restrictive kind of connectivity that for any two points requires the existence of a continuous mapping from $\mathbb{R}$ that passes through them, i.e. there is a path connecting any two points
- **Homeomorphism**: a continuous bijective mapping with a continuous inverse; the most basic equivalency in topology

A **topological n-manifold** (AKA n-manifold) is then defined to be a Hausdorff space in which every point has an open neighborhood homeomorphic to an open subset of $\mathbb{R}^n$. A **topological n-manifold with boundary** allows the neighborhoods of points to be homeomorphic to an open subset of the closed half of $\mathbb{R}^n$, i.e. of the portion of $\mathbb{R}^n$ on one side of and including $\mathbb{R}^{n-1}$. The points that map to $\mathbb{R}^{n-1}$ form the **boundary** of the manifold, and are a $(n-1)$-manifold. A compact manifold without boundary is often called a **closed manifold** to distinguish it from a compact manifold with boundary. Most definitions of a manifold usually also include some additional technical requirement to avoid pathological exceptions (e.g. second countable $\Rightarrow$ paracompact $\Leftrightarrow$ metrizable, none of which we will define here).

Topological spaces include many interesting objects that do not lie on our path towards manifolds, for example objects constructed by taking the limit of some iterative process. In this book we will be mainly interested in results that are valid for spaces with the "nicer" manifold structure. For example, for manifolds the definitions of path-connected and connected are equivalent.

4.2.2 Mapping spaces

For algebraic objects, the most basic structure-preserving map was a homomorphism. The most basic equivalence in topology is the similarly named homeomorphism. Homeomorphic spaces are "the same" from a topological point of view.

☼ One can visualize a homeomorphism as stretching and bending a space arbitrarily, since length and curvature involve structure beyond open sets and so are "invisible" from the topological viewpoint.

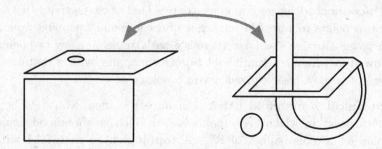

Figure 4.2.1 Visualizing homeomorphic spaces.

There exists an even looser equivalency in topology, called a **homotopy equivalency**. The essential conceptual difference is that since it is bijective, a homeomorphism preserves the dimensionality of the object while stretching and bending, while a homotopy equivalency allows the collapse of dimensions, while still preserving holes.

To make this concept precise, we first define a **homotopy** between spaces, a continuous family of continuous maps $f_t \colon X \to Y$; i.e. f is continuous when considered as a function of t as well as when considered as a function of points in X. Two maps are **homotopic** if there is a homotopy between them, i.e. f_0 is homotopic to f_1 if t runs from 0 to 1.

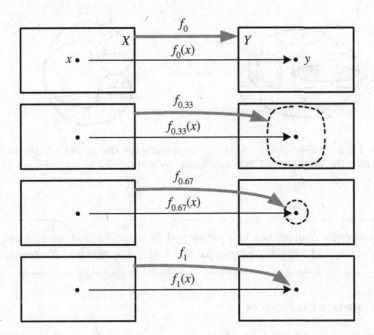

Figure 4.2.2 A homotopy f_t from X to Y; the map f_0 onto Y is homotopic to the map f_1 which maps all of X to a point y.

A couple of related definitions are:

- **Homotopy relative** to A (denoted homotopy rel A): a homotopy that is independent of t on $A \subset X$; e.g. in the above figure f_t is a homotopy rel x
- **Deformation retraction** from X to A: a homotopy rel A from the identity to a retraction from X to A; e.g. if we take $Y = X$ and $y = x$ above, f_t is a deformation retraction from X to x

The precise definition of two spaces being homotopy equivalent, or having the same **homotopy type** (denoted $X \simeq Y$), is not important for our purposes; instead, we will state the derived fact that $X \simeq Y$ iff they are deformation retracts of the same space. A space homotopy equivalent to a point is then called **contractible**.

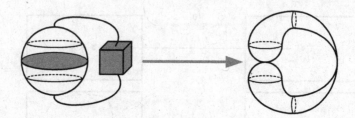

Figure 4.2.3 A deformation retraction that collapses the equatorial disc inside the sphere, the strings, and the solid cube to points; the two spaces are thus homotopy equivalent.

☼ Homotopy equivalency can be viewed as meaning that we can collapse or expand any parts of the space as well as bending and stretching.

4.3   Constructing spaces

Spaces are often defined by taking a geometric view of non-geometric objects, as in spaces of functions or spaces of solutions to a differential equation. However, one can also construct spaces from scratch.

4.3.1   *Cell complexes*

One important way to construct spaces is using the procedure that defines a **cell complex** (AKA CW-complex):

Start with a discrete set of points, called **0-cells**:

●   ●   ●

Attach line segments, called **1-cells**, by their boundaries to form the **1-skeleton**:

   ●

Attach disks, called **2-cells**, by their boundaries to form the **2-skeleton**:

We can then continue to any dimension by iteratively attaching **n-cells**, n-dimensional disks D^n in $\mathbb{R}^n$, by their boundaries, which are $(n-1)$-dimensional spheres S^{n-1}, to form the **n-skeleton**.

The first construction above illustrates that a torus with g holes (**genus g**) can be built from one 0-cell, $2g$ 1-cells, and one 2-cell; the second illustrates that a sphere S^n is one n-cell attached to one 0-cell; and the third illustrates that the **characteristic maps** by which cells are attached can change the homotopy type of the resulting cell complex (in this case to an object which is not homotopy equivalent to the torus).

Can we always find a cell complex structure for the spaces we will be concerned with, mainly manifolds? Some facts are:

- Every compact manifold is homotopy equivalent to a cell complex
- Closed manifolds can always be given a cell complex structure, except in dimension 4 where the answer is unknown

4.3.2 Projective spaces

We can also define spaces in other ways, and then try to find cell complex structures for them. For example, the **real projective n-space** $\mathbb{RP}^n$ is defined as the space of all lines through the origin in $\mathbb{R}^{n+1}$. Each such line is determined by a unit vector, except that the negative of every vector is identified with the same line, so we can consider $\mathbb{RP}^n$ to be S^n with antipodal points identified.

Alternatively, we can look at $\mathbb{RP}^n$ as the unit vectors in the upper hemisphere only, since the lower hemisphere is made up of all negatives of the upper; except that now, antipodal points of the boundary are identified. But the upper hemisphere is D^n and its boundary is S^{n-1} with antipodal points identified, or $\mathbb{RP}^{n-1}$. Thus $\mathbb{RP}^n$ is obtained by attaching an n-cell to $\mathbb{RP}^{n-1}$, and by induction we can see that $\mathbb{RP}^n$ has a cell complex structure with one cell in each dimension up to n.

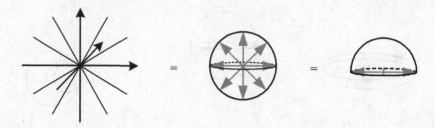

Figure 4.3.1 Different ways of viewing the real projective 2-space $\mathbb{RP}^2$.

The identifications in the above constructions of $\mathbb{RP}^2$ are not easily visualized, since they cannot be embedded in $\mathbb{R}^3$; in contrast, $\mathbb{RP}^1$ is D^1 with boundary S^0 having antipodal points identified, i.e. the line segment with the endpoints identified; in other words $\mathbb{RP}^1 = S^1$, the circle.

We can also define the **complex projective n-space** $\mathbb{CP}^n$, which is the space of all lines through the origin in $\mathbb{C}^{n+1}$. In this case one has a cell complex structure with one cell in each even dimension up to $2n$. $\mathbb{HP}^n$ can similarly be defined, but $\mathbb{OP}^n$ can only be defined for $n < 3$ due to lack of associativity. By generalizing the reasoning above, we have $\mathbb{CP}^1 = S^2$; $\mathbb{HP}^1 = S^4$; and $\mathbb{OP}^1 = S^8$. As manifolds, the projective spaces are all closed, i.e. compact and without boundary.

4.3.3 Combining spaces

Another way to construct spaces is by combining them in various ways, as described in the following table. We denote the unit interval $[0, 1]$ by I. $X - Y$ simply denotes the usual removal of a subset Y.

Table 4.3.1 Operations on spaces.

Operation	Example
Product: $X \times Y =$ all points (x, y) with $x \in X$, $y \in Y$ Example: $S^1 \times S^1 = T^2$, the torus	
Wedge sum: $X \vee Y$ identifies a point from each space Example: $S^1 \vee S^1 =$ the figure eight	
Quotient: X/A collapses $A \subset X$ to a point Example: $S^2/S^1 = S^2 \vee S^2$	
Suspension: $SX =$ the quotient of $X \times I$ with $X \times \{0\}$ and $X \times \{1\}$ collapsed to points Example: $SS^1 = S^2$	
Join: $X * Y =$ all line segments from X to Y; more precisely, $X \times Y \times I$ with $X \times Y \times \{0\}$ collapsed to X, $X \times Y \times \{1\}$ to Y Example: $I * I =$ solid tetrahedron	

In the case of two disjoint connected n-manifolds X and Y, we can also define the **connected sum** $X \# Y$, obtained by removing the interiors of closed n-balls from each and identifying the resulting boundary spheres.

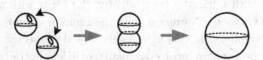

Figure 4.3.2 The connected sum of two manifolds removes a ball from each and identifies their boundaries; in this example we have $S^2 \# S^2 = S^2$.

4.3.4   *Classifying spaces*

Our real interest is in manifolds, and it is their classification that we will discuss in this section. In addition, we will refer to the concept of orientability here, which is defined for topological spaces in the next chapter.

As intuitive 2-dimensional guides, we have now utilized the sphere S^2, the torus T^2, and projective plane $\mathbb{RP}^2$. In fact, this essentially exhausts the topology of closed connected 2-manifolds, a full classification of which is:

- Orientable: the sphere, the torus, or the connected sum of tori
- Non-orientable: the projective plane, or a connected sum of projective planes

The Klein bottle, another well-known non-orientable surface (see Figure 5.1.3), is homeomorphic to the connected sum $\mathbb{RP}^2 \# \mathbb{RP}^2$. Any compact manifold with boundary is then obtained from these spaces by removing one or more open disks.

Another way of describing this classification scheme is using handles. A three-dimensional **handlebody** is a ball with g handles attached to it, which is equivalent to the solid torus of genus g. Attaching a handle also refers to the corresponding operation on the boundary, i.e. adding a cylinder to a sphere by removing two disks. The concept of adding handles to a manifold is generalized in **surgery theory**.

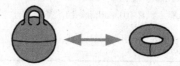

Figure 4.3.3 The ball with one handle is homeomorphic to the solid torus; from the point of view of boundaries, a sphere with one handle is homeomorphic to the torus.

The sphere with g handles (equivalently, the torus with g holes) is referred to as the **surface of genus g**, with the genus of a handlebody defined by that of its boundary. In this light, every closed connected 2-manifold is homeomorphic to a sphere, projective plane, or Klein bottle, possibly with some number of attached handles.

This simple classification scheme does not extend to higher dimensions, a clear indication of the limits of intuition in higher dimensional spaces.

However, the following additional facts can be stated regarding the classification of manifolds:

- Every connected 1-manifold without boundary is homeomorphic to either the line $\mathbb{R}$ or the circle S^1
- The classification of 3-manifolds was accomplished by the 2003 proof of Thurston's geometrization conjecture
- It can be proved that the classification of manifolds of dimension four or greater is impossible
- Many properties can be proved for dimension five or greater, but not for four dimensions; for example, the simply-connected manifolds of dimension five or greater are completely classified

The recent classification of 3-manifolds is of particular interest, since it also proved the long-standing **Poincaré conjecture**, which generalized to arbitrary dimension states that every closed n-manifold that is homotopy equivalent to the n-sphere is homeomorphic to the n-sphere. Since 1904 when the conjecture was first stated, it has been proven for every dimension except 3, with dimension 4 having been solved in 1982. Every compact orientable 3-manifold can be obtained by identifying the boundaries of two handlebodies of the same genus in some way; this is called a **Heegaard splitting** (AKA Heegaard decomposition). This does not amount to a classification, however, since different splittings can yield the same manifold.

Chapter 5

Algebraic topology

Algebraic topology is concerned with characterizing spaces. The main tools used to do this, called homotopy groups and homology groups, measure the "holes" of a space, and so are invariant under homotopy equivalence. More precisely, these objects are functors from the category of spaces and continuous maps to that of groups and homomorphisms.

One might hope that some definitive conclusion could be reached if all of these objects were isomorphic for a given space, but unfortunately this is not the case: different spaces may have identical measurements.

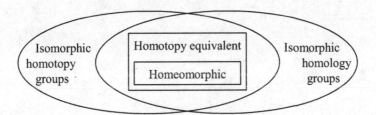

Figure 5.0.1 Algebraic topology differentiates spaces better than proving they are the same.

Thus the tools of algebraic topology are best viewed as incomplete measurements, or "shadows" of a space: if they are different, we know the spaces are distinct, but if they are the same, we cannot conclude anything in general.

5.1   Constructing surfaces within a space

5.1.1   *Simplices*

As a tool for measuring "holes" in the next section, we will need to construct n-dimensional surfaces within a space. We will build these surfaces out of simple triangular pieces called **simplices**. The idea is to map triangles to the space, then build surfaces out of these triangles within the space.

- **Standard n-simplex Δ^n**: the n-dimensional space defined by the join of the $(n+1)$ unit points in $\mathbb{R}^{n+1}$
- **Singular n-simplex** (AKA n-simplex): a mapping $\sigma : \Delta^n \to X$ of the standard n-simplex into X, the space being studied; the map is only assumed to be continuous (and so may be singular)
- **n-chain**: a finite formal linear sum ("chain") $\sum a_\alpha \sigma_\alpha$, where σ_α are n-simplices and coefficients a_α are integers; under addition of sums the n-chains form an abelian group denoted $C_n(X)$
- **Boundary homomorphisms $\partial_n : C_n(X) \to C_{n-1}(X)$**: takes an n-chain to the $(n-1)$-chain consisting of the oriented sum of boundaries; also denoted simply ∂

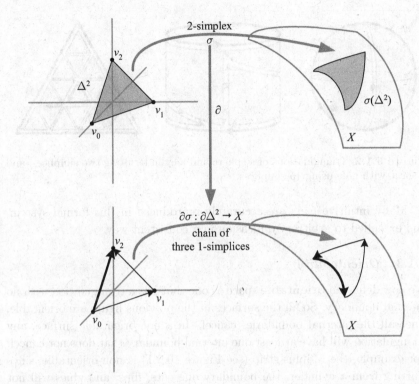

Figure 5.1.1 The boundary homomorphism of a singular 2-simplex to three 1-simplices.

By oriented sum, it is meant that we desire a coherent orientation in X; i.e. the boundary of a boundary should vanish. We can achieve this in any dimension by taking the boundaries $\partial \Delta^n$ as oriented according to vertex order, and then defining $\partial \sigma \left(\partial \Delta^n \right) \equiv \sum \left(-1 \right)^i \sigma \left[\Delta^n - v_i \right]$. This formalism has the effect of reversing the "backwards" boundaries by preceding them with a minus sign, as can be seen with the heavier weight arrows in the preceding figure for a 2-simplex.

5.1.2 *Triangulations*

The point of introducing orientation is to use n-chains to describe arbitrary n-dimensional surfaces in X composed of n-simplices. By constructing such surfaces using adjacent simplices, internal boundaries can cancel when they consist of two boundaries in opposite directions. Constructing a surface out of simplices in this way is called a **triangulation**.

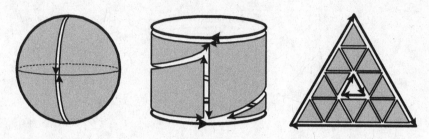

Figure 5.1.2 Triangulations of a sphere and a cylinder using two simplices, and a sheet with hole using 15 simplices.

Many intuitive facts are accurately reproduced in this formal system, and extended to arbitrary dimension in a consistent way.

5.1.3 *Orientability*

We now define an **orientable** space as one that has a triangulation with no internal boundary. So all the surfaces in the previous figure are orientable, since all the internal boundaries cancel. In a non-orientable surface, any triangulation will have at least one internal boundary that does not cancel. For example, the Möbius strip (see Figure 10.2.1) is non-orientable, since starting from a cylinder, the boundary one cuts, flips, and glues will not cancel.

The simplest closed non-orientable surface is the **Klein bottle**. We can see from the preceding figure that a triangulated cylinder has external boundaries oppositely oriented. Thus in order to cancel internal boundaries when forming a closed surface out of cylinders, each new cylinder must begin with a boundary of the same orientation. In the following figure we illustrate building the Klein bottle out of cylinders, noting that upon return to the "mouth" at left from the "inside," we are faced with a non-canceling internal boundary.

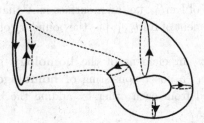

Figure 5.1.3 A triangulation of the Klein bottle will have a non-canceling boundary, as seen here by following the leading edge of an oriented cylinder around the bottle.

The other familiar non-orientable space we've encountered is $\mathbb{RP}^2$. In general, $\mathbb{RP}^n$ is orientable if n is odd, and is non-orientable if n is even.

5.1.4 *Chain complexes*

A fundamental intuitive fact reproduced in this formalism is that the boundary of a boundary is zero. A useful algebraic generalization of this idea is the **chain complex**, defined to be a sequence of homomorphisms of abelian groups $\partial_n \colon C_n \to C_{n-1}$ with $\partial_n \partial_{n+1} = 0$. In our case the abelian groups C_n are the n-chains, and the chain complex can be illustrated as follows:

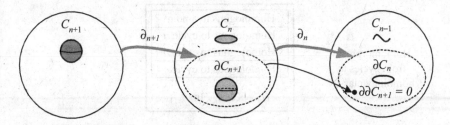

Figure 5.1.4 The chain complex of n-chains of simplices via boundary homomorphisms. The figures are representative of general n-chains, which of course can be much more complicated.

5.2 Counting holes that aren't boundaries

5.2.1 *The homology groups*

We can now define the homology groups (AKA singular homology groups). In the below definitions, Ker denotes kernel and Im denotes image.

- **Cycle**: an element of $\mathrm{Ker}\partial_n$, i.e. an n-surface in X that has no boundary
- **Boundary**: an element of $\mathrm{Im}\partial_{n+1}$, i.e. the boundary of an $(n+1)$-volume in X
- **Homology class**: an element of the **homology group** $H_n(X) \equiv \mathrm{Ker}\partial_n/\mathrm{Im}\partial_{n+1}$, i.e. a coset consisting of **homologous** n-cycles that can all be obtained from each other by adding the boundary of some $(n+1)$-volume in X

> ✿ We can think of a typical cycle as a loop for $n = 1$ or a sphere for $n = 2$. Then the typical boundary is the chain of n-cycles that form the edge of an arbitrary surface for $n = 1$ or the surface of an arbitrary volume for $n = 2$.

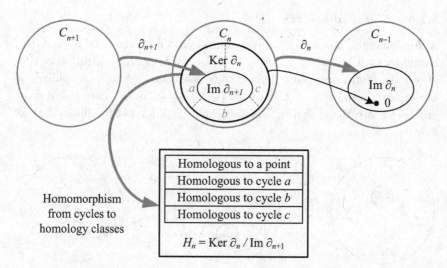

Figure 5.2.1 Each element of H_n consists of cycles whose difference is a boundary.

The cylinder and punctured plane in Figure 5.1.2 depict examples of homologous loops, two 1-chains that are the boundary of a 2-chain. The abelian group $H_n(X)$ is then generated by the cosets of non-homologous n-cycles, thus counting the number of "n-dimensional holes" in X.

5.2.2   Examples

The construction of the homology groups is somewhat complicated, but the idea behind it is quite intuitive. For simple spaces, we expect that H_n will be a direct product of $\mathbb{Z}$ components, one for each "n-dimensional hole" in X that is not the boundary of a $(n+1)$-volume, since each such hole can be wrapped around any number of times in either direction, and none of these wrappings are homologous.

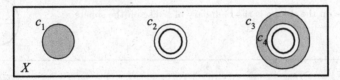

Figure 5.2.2 The cycle c_1 is the boundary of a disc, and so is homologous to a point. A hole in X prevents c_2 from being the boundary of any surface. c_3 is homologous to c_4 since their difference is the boundary of an annulus, thus preventing the hole from being counted twice. Thus $H_1(X) = \mathbb{Z} \oplus \mathbb{Z}$. Note that a cycle around both holes (not depicted) would be homologous to the cycle $c_2 + c_4$.

The best way to see this is to consider some examples.

Table 5.2.1 Homology groups of simple geometric shapes.

Homology Group	Diagram
$H_1(S^1) = \mathbb{Z}$ A loop can be mapped to a circle by wrapping around any number of times in either direction	
$H_1(S^2) = 0$ Any circle on the sphere is the boundary of a disc on the sphere	
$H_2(S^1) = 0$ Any sphere mapped along the edge of a circle is always the boundary of a ball also mapped along the circle	
$H_1(T^2) = \mathbb{Z} \oplus \mathbb{Z}$ The two circles that make up the 1-skeleton of a torus are not homologous and thus a loop can wrap around either circle any number of times in either direction	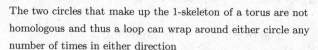
$H_2(T^2) = \mathbb{Z}$ A torus can be mapped to itself by wrapping around any number of times in either direction	

These calculations reflect the close relationship between homology groups and cell complex structure. Since the nth homology group is defined in terms of n-surfaces, we intuitively expect $H_n(X)$ to only depend on the $(n+1)$-skeleton of X (which recall includes all k-cells for $k \leq n+1$), and this is in fact true if X is a cell complex. Thus if X is a cell complex with finitely many cells, for example a closed manifold, $H_n(X)$ is a finitely generated group.

For example, inductively extending the first three observations to spheres in arbitrary dimension shows that $H_n(S^d) = \mathbb{Z}$ if $n = d$, 0 otherwise. For the d-dimensional torus T^d, $H_n(T^d)$ is the direct sum of c copies of $\mathbb{Z}$ with c the binomial coefficient d choose n.

5.2.3   Calculating homology groups

It should be kept in mind that although homology, like most of algebraic topology, is geometrically inspired, its algebraic constructions may or may not have ready geometric interpretations in odd situations or higher dimensions. Despite the "sensible" results above, counter-intuitive facts caution us to always deduce these and other measurements of a space with mathematical rigor. For example, the "hole" interpretation of homology does not easily extend to non-orientable surfaces:

- $H_1\left(\mathbb{RP}^2\right) = \mathbb{Z}_2$, since if we view $\mathbb{RP}^2$ as D^2 with boundary S^1 having antipodal points identified, the path connecting two antipodal points is a "loop," but any two such loops together is homologous to a point
- $H_2\left(\mathbb{RP}^2\right) = 0$, since any triangulation of $\mathbb{RP}^2$ has a boundary, the real projective plane being non-orientable
- In general, any non-orientable manifold M^n has $H_n(M^n) = 0$

As one might guess from the examples considered so far, it is also a fact that the topology of 2-manifolds is completely determined by homology. This circumstance is certainly not true in higher dimensions, as we noted in the introduction to this chapter.

There are various relations that can help in calculating homology groups. For example, an immediate result is that if X has connected components X_i, $H_n(X) = \bigoplus_i H_n(X_i)$. Another is the **excision theorem**, which states that for $Z \subset A \subset X$, $H_n(X - Z, A - Z) \cong H_n(X, A)$. Here $H_n(X, A)$ is a **relative homology group**, defined using n-chains $C_n(X, A) \equiv C_n(X)/C_n(A)$ in place of $C_n(X)$; this construction essentially ignores any holes in $A \subset X$. Thus the excision theorem states the intuitively expected fact that we can delete any portion of a subspace A without affecting $H_n(X, A)$, which already ignores holes in A.

5.2.4   Related constructions and facts

There are many homological tools used in algebraic topology. Some variants of the homology groups include:

- **Homology group with coefficients**: for an abelian group G, $H_n(X; G)$ is defined using n-chains $C_n(X; G)$ with coefficients in G instead of $\mathbb{Z}$
- **Reduced homology groups**: a slight variant that avoids the result $H_0 = \mathbb{Z}$ for points while keeping the higher homology groups the same

- **Local homology** of X at A: $H_n(X \mid A) \equiv H_n(X, X - A)$ depends only on a neighborhood of A in X
- **Simplicial homology, cellular homology**, etc.: more easily constructed homology theories that are only valid for certain types of spaces; all can be shown to be equivalent to singular homology for those spaces
- **Cohomology groups**: $H^n(X; G)$ are dual constructions based on the **cochain groups** $C^n(X; G) \equiv C_n^* = \operatorname{Hom}(C_n, G)$, the group of homomorphisms from C_n to some abelian group G; a homomorphism $H^n(X; G) \to \operatorname{Hom}(H_n(X; G), G)$ can be constructed which is surjective, becoming an isomorphism if G is a field
- **Cohomology ring**: $H^*(X; R)$ is a direct sum of the cohomology groups $H^n(X; R)$ with coefficients in a ring R; multiplication is defined using the **cup product**, a product between the $H^n(X; R)$

Some related constructions include:

- **Betti number**: $b^n \equiv$ the number of $\mathbb{Z}$ summands if $H_n(X)$ is written $\mathbb{Z} \oplus \cdots \oplus \mathbb{Z}_{c_1} \oplus \mathbb{Z}_{c_2} \oplus \mathbb{Z}_{c_3} \oplus \cdots$, where the c_i are called **torsion coefficients**; in a cell complex, b^n is just the number of n-cells
- **Euler characteristic**: the alternating sum of Betti numbers $\chi = b^0 - b^1 + b^2 - \cdots$, i.e. for a cell complex the number of even cells minus the number of odd cells; thus a compact connected surface has genus $g = (2 - \chi)/2$
- **Brouwer degree** (AKA winding number for S^1): any mapping $\phi \colon S^n \to S^n$ induces a homomorphism on $H_n(S^n) = \mathbb{Z}$ of the form $z \to az$; the integer a is the Brouwer degree of the map, essentially the number of times the mapping wraps around the sphere
- **Moore space**: given an abelian group G and an integer $n > 0$, the space $M(G, n)$ is constructed to have $H_n = G$ and $H_i = 0$ for $i \neq n$

5.3 Counting the ways a sphere maps to a space

Besides homology, the other major measuring tool in algebraic topology is that of the homotopy groups $\pi_n(X)$. These functors count the number of non-homotopic maps existing from S^n to the space X. Recalling the definition of homotopic from Section 4.2.2 on page 58, we see that this is essentially a count of the classes of mapped spheres that cannot be deformed into each other.

Although the construction of the homotopy groups is much simpler than that of the homology groups, what they actually measure in the space is less easily described in an intuitive way, for example in terms of "holes."

This is apparent from the fact that like the homology groups, $\pi_i(S^n) = \mathbb{Z}$ for $i = n$ and vanishes for $i < n$; however for $i > n$ we have $H_i(S^n) = 0$ as we would intuitively expect, while $\pi_i(S^n)$ yields a complicated table of groups that comprises an area of active research. This is representative of the fact that the homotopy groups are in general much harder to compute than the homology groups.

5.3.1 The fundamental group .

The simplest homotopy group is the **fundamental group** $\pi_1(X)$, which counts how many ways a loop can be mapped into a path-connected space X. More precisely, we define $\pi_1(X, x)$ to be the set of all homotopy classes of parameterized loop mappings that begin and end at a **basepoint** x. For a path-connected space we can add a path from x to any other point and back as part of the loop, so that $\pi_1(X, x)$ is independent of x and is written $\pi_1(X)$.

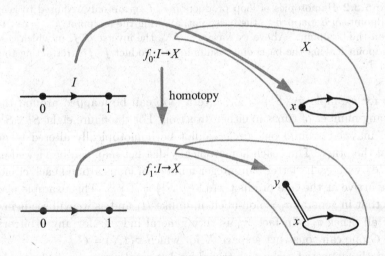

Figure 5.3.1 For a path-connected space X, every loop with basepoint x is homotopic to a loop with basepoint y, so that $\pi_1(X, x) = \pi_1(X, y) = \pi_1(X)$.

$\pi_1(X)$ becomes a group by defining multiplication as the composition of loop paths from a given basepoint. Technically, this is implemented by dividing the line segment I in half, and applying the first mapping f to the first half segment, g to the second half. Note that this means that homotopies of fg may abandon the midway mapping to the basepoint; for example, the inverse of a path is simply the same path traversed backwards.

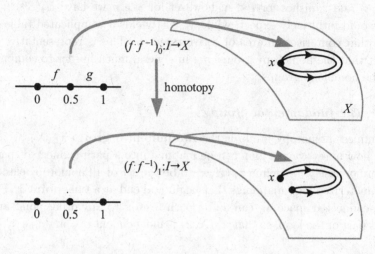

Figure 5.3.2 Homotopies of loop products $h = f \cdot g$ are only required to keep the endpoints of h mapped to the basepoint, allowing the midpoint $f_{0.5} = g_{0.5}$ to abandon the basepoint. Above we have $g = f^{-1}$, the inverse of f, in which case the midpoint leaving the basepoint results in the product $f \cdot f^{-1}$ retracting to a point.

So for example $\pi_1 \left(S^1 \right) = \mathbb{Z}$, since a loop can be mapped around the circle any number of times in either direction. For the figure eight $S^1 \vee S^1$, a loop mapped around one circle cannot be homotopically altered to go around the other. Thus each homotopically distinct map can be viewed as a "word" with each "letter" an integer number of loops around each circle, i.e. we arrive at the free product $\pi_1 \left(S^1 \vee S^1 \right) = \mathbb{Z} * \mathbb{Z}$. This example also shows that in general π_1 is non-abelian, unlike H_1 and as we will see below unlike all other π_n. In fact, π_1 is very general indeed: for any arbitrary group G one can construct a space X for which $\pi_1 (X) = G$.

A path-connected space with trivial fundamental group is called **simply connected**. The name reflects the fact that in such a space there is only one homotopically distinct way to form a path between any two points.

$\triangle$ Note that there are other definitions of "simply connected" in use; some do not require the space to be connected.

5.3.2 The higher homotopy groups

Generalizing the fundamental group to higher dimensions yields the **homotopy groups** $\pi_n(X)$, which count how many ways S^n can be mapped into a path-connected space X.

Group multiplication is similarly implemented by dividing the n-cube I^n in half along one dimension, and applying the maps to be multiplied to each half. Just as we required homotopies of loops to map endpoints to the basepoint, we here require the boundary of I^n to map to the basepoint. Unlike π_1, π_n is abelian for $n > 1$. This is because there is no "room" in one dimension to homotopically swap the spheres as there is in two or more dimensions.

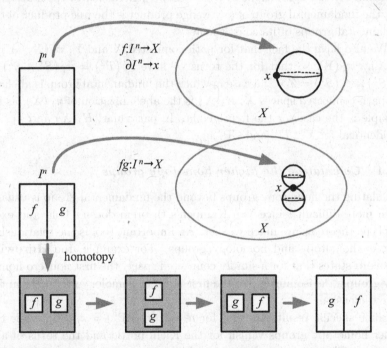

Figure 5.3.3 In more than one dimension, maps from the two halves of I^n can be homotopically swapped, making π_n abelian. The shaded surfaces in the figure are mapped to the basepoint x along with the boundaries.

Although the homotopy groups as a measuring tool share the incompleteness that characterizes all of algebraic topology, i.e. equal π_n do not guarantee homotopy equivalent spaces, there is a theorem that comes close.

Whitehead's theorem states that a map between cell complexes that induces isomorphisms on all π_n is a homotopy.

5.3.3 Calculating the fundamental group

There are several tools that facilitate calculating the fundamental group. **Van Kampen's theorem** can be used to compute the fundamental group of a space in terms of simpler spaces it is constructed from. If certain conditions are met, the theorem states that for $X = \bigcup A_\alpha$, $\pi_1(X) = \underset{\alpha}{*}\pi_1(A_\alpha)$, the free product of the component fundamental groups. Under less restrictive conditions this becomes a factor group of the free product. For example the result for the figure eight $S^1 \vee S^1$ can be generalized to the statement that the fundamental group of any wedge product is the free product of the fundamental groups of its constituents.

We also have the fact that for path-connected X and Y, $\pi_1(X \times Y) = \pi_1(X) \times \pi_1(Y)$, so that for the torus we have $\pi_1(T^2) = \pi_1(S^1 \times S^1) = \pi_1(S^1) \times \pi_1(S^1) = \mathbb{Z} \times \mathbb{Z}$, a case in which the fundamental group is abelian. For path-connected spaces X, $H_1(X)$ is the abelianization of $\pi_1(X)$, as for example is the case for the figure eight. In particular, $H_1(X)$ and $\pi_1(X)$ are identical for S^n, T^n, and $\mathbb{C}\mathrm{P}^n$.

5.3.4 Calculating the higher homotopy groups

Calculating the homotopy groups beyond the fundamental group is usually much more difficult, since Van Kampen's theorem doesn't hold and excision type theorems are much weaker. An important tool is the relationship between homotopy and homology groups. For example, the **Hurewicz theorem** states that for a simply connected space, the first nonzero homotopy group π_n is isomorphic to the first nonzero homology group H_n in the same dimension.

Some specific results are that for $n > 1$, $\pi_n(\mathbb{R}\mathrm{P}^d) = \pi_n(S^d)$, while the higher homotopy groups vanish for the Klein bottle and the torus of any dimension T^n.

5.3.5 Related constructions and facts

Some constructions related to the homotopy groups include:

- **Relative homotopy groups**: $\pi_n(X, A, x_0)$ is defined to be all homotopy classes of maps $(D^n, \partial D^n, s_0) \to (X, A, x_0)$ where $s_0 \in S^{n-1}$ and $A \subset X$

- **Stable homotopy group:** $\pi_i^s(X)$ is defined to be the group that the sequence $\pi_i(X) \to \pi_{i+1}(SX) \to \pi_{i+2}(S^2 X) \to \cdots$ eventually arrives at, recalling from Section 4.3.3 the suspension SX; a major unsolved problem in algebraic topology is computing the stable homotopy groups of the spheres

- **n-connected space:** indicates that $\pi_i = 0$ for $i \le n$; so 0-connected = path-connected, 1-connected = simply connected

- **Action of π_1 on π_n:** the homomorphism $\pi_1 \to \mathrm{Aut}\,(\pi_n)$ defined by taking the basepoint of π_n around the loop defined by each element of π_1

- **n-simple space:** indicates trivial action of π_1 on π_n; a **simple** (AKA abelian) space is n-simple for all $n \Rightarrow \pi_1$ is abelian

- **Eilenberg-MacLane space:** a space $K(G,n)$ constructed to have one nontrivial $\pi_n = G$, a rare case of π_n uniquely determining the homotopy type; for any connected cell complex X, one can construct a $K(G,n)$ such that the homotopy classes of maps from X to $K(G,n)$ are isomorphic to $H^n(X; G)$, thus turning homology groups into homotopy groups

Chapter 6

Manifolds

In studying spaces, we considered the locally Euclidean structure of topological manifolds as defining a subset of spaces that were "nicer," meeting the minimum requirements of our idea of a geometrical "shape" such as integral dimension. By slightly narrowing our consideration to differentiable manifolds, we can essentially graft calculus onto our "rubber sheet." The constructions of coordinates and tangent vectors enable us to define a family of derivatives associated with the concept of how vector fields change on the manifold. The challenge is in defining all these objects without an ambient space, which our intuitive picture normally depends upon.

> △ Note that a differentiable manifold includes no concept of length or distance (a metric), and no structure that allows tangent vectors at different points to be compared or related to each other (a connection). It is important to remember that nothing in this chapter depends upon these two extra structures.

When dealing with manifolds, there are two main approaches one can take: express everything in terms of coordinates, or strive to express everything in a coordinate-free fashion. In keeping with this book's attempt to focus on concepts rather than calculations, we will take the latter approach, but will take pains to carefully express fundamental concepts in terms of coordinates in order to derive a picture of what these coordinate-free tools do. Facility in moving back and forth between these two views is a worthwhile goal, best accomplished by combining the material here with a standard text on differential geometry or general relativity.

6.1 Defining coordinates and tangents

6.1.1 *Coordinates*

Recall that the key feature of a topological manifold M^n is that every point has an open neighborhood homeomorphic to an open subset of $\mathbb{R}^n$. To make this precise we define the following terms.

- **Coordinate chart** (AKA parameterization, patch, system of coordinates): a homeomorphism $\alpha\colon U \to \mathbb{R}^n$ from an open set $U \subset M^n$ to an open subset of $\mathbb{R}^n$
- **Coordinate functions** (AKA coordinates): the maps $a^\mu\colon U \to \mathbb{R}$ that project α down to one of the canonical Cartesian components
- **Atlas**: a collection of coordinate charts that cover the manifold
- **Coordinate transformation** (AKA change of coordinates, transition function): in a region covered by two charts, we can construct the map $\alpha_2 \circ \alpha_1^{-1}\colon \mathbb{R}^n \to \mathbb{R}^n$

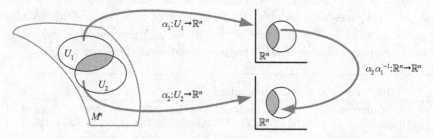

Figure 6.1.1 In the intersection of two coordinate charts we can construct the coordinate transformation, a homeomorphism on $\mathbb{R}^n$.

$\triangle$ A coordinate chart is sometimes defined to be the inverse map $\alpha^{-1}\colon \mathbb{R}^n \to M$ valid on an open subset of $\mathbb{R}^n$, with similar changes to related definitions such as coordinate functions.

The coordinate transformations are simply maps on Euclidean space, so we can require them to be infinitely differentiable (AKA smooth, C^∞). An atlas whose charts all have smooth coordinate transformations determines a **differentiable structure**, which turns the topological manifold into a

differentiable manifold (AKA smooth manifold). Two differential structures are considered to be equivalent if the union of their atlases still results in smooth coordinate transformations. Unless otherwise noted, from this point forward "manifold" will mean differentiable manifold.

A **complex manifold** is defined to have an atlas of charts to $\mathbb{C}^n$ whose coordinate transformations are analytic. Complex n-manifolds are a subset of real $2n$-manifolds, but atlases are highly constrained since complex analytic functions are much more constrained than smooth functions. By "manifold" we will always mean a real manifold in this book.

With the addition of a differentiable structure, one can define the various tools of calculus on manifolds in a straightforward way. Differentiable functions $f \colon U \to \mathbb{R}$ require the map $f \circ \alpha^{-1} \colon \mathbb{R} \to \mathbb{R}$ to be differentiable, and differentials $\partial/\partial a^\mu$ are defined at a point $p \in U$ by

$$\partial/\partial a^\mu \mid_p f \equiv \partial/\partial x^\mu \left(f \circ \alpha^{-1}(x) \right) \mid_{x=\alpha(p)} .$$

where $x \in \mathbb{R}^n$. All of the usual relations of calculus hold with these definitions.

$\triangle$ To avoid clutter, a common abuse of notation is to use x^μ to denote any or all of three quantities: the point $p \in M$, the coordinate functions $a^\mu \colon M \to \mathbb{R}$, and the $\mathbb{R}^n$ n-tuplet $x^\mu = a^\mu(p)$. Similarly, the differential $\partial/\partial a^\mu$ is often denoted $\partial/\partial x^u$. We will follow these conventions going forward, but when dealing with fundamental definitions or pictures, it is important to distinguish these very different quantities from each other. Another shortcut is to denote differentials by ∂_μ; as with basis vectors, it is important to remember that these are labels, not component indices.

6.1.2 Tangent vectors and differential forms

The **tangent space** T_pU at a point $p \in U$ is defined to be the vector space spanned by the differential operators $\partial/\partial a^\mu \mid_p$. A **tangent vector** $v \in T_pU$ can then be expressed in tensor component notation as $v = v^\mu \partial/\partial a^\mu$, so that $v(a^\mu) = v^\mu$. The tangent vector $\partial/\partial a^\mu \mid_p$ applied to a function f can be thought of as "the change of f in the direction of the μth coordinate line at p."

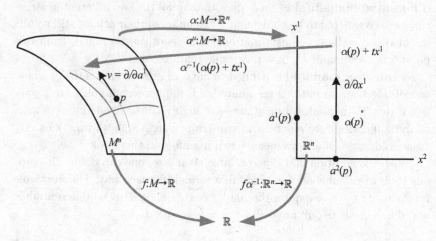

Figure 6.1.2 In a particular coordinate chart, a tangent vector v operates on a function by taking the derivative of the composite function in $\mathbb{R}^n$ in the direction of $v^\mu \partial / \partial a^\mu$.

Thus at a point p, we have

$$v^\mu \frac{\partial}{\partial a^\mu} (f) = v^\mu \frac{\partial}{\partial x^\mu} \left(f \circ \alpha^{-1} (x) \right),$$

where $x = \alpha(p)$. The coordinate line $\alpha^{-1} \left(a^\mu (p) + t v^\mu x^\mu \right)$ is a parameterized curve on M, and thus it and the tangent vector itself are coordinate-independent objects. In another coordinate chart, the coordinate line that yields the same operator on functions near p can be seen to correspond to the familiar transformation of vector components

$$v = v^\mu \frac{\partial}{\partial a^\mu} = \left(v^\lambda \frac{\partial b^\mu}{\partial a^\lambda} \right) \frac{\partial}{\partial b^\mu}.$$

We can consider the point "p moved in the direction v" by abusing notation to write $p^\mu + t v^\mu$ in place of $\alpha^{-1} \left(a^\mu (p) + t v^\mu x^\mu \right)$; this is a coordinate-dependent expression, but in the limit $\varepsilon \to 0$ we can unambiguously write $p + \varepsilon v$ to refer to the concept "p moved infinitesimally in the direction v," which is coordinate-independent.

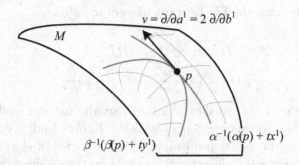

Figure 6.1.3 A tangent vector v in terms of two different coordinate charts. $v^\mu = (1,0)$ in chart α with coordinate functions $a^\mu(p) = x^\mu$, and $v^\mu = (2,0)$ in chart β with coordinate functions $b^\mu(p) = y^\mu$. The divergent coordinate lines show that the concept of moving a point "in the direction of v" can only be coordinate-independent in the infinitesimal limit.

The set of all tangent spaces in a region U is called the **tangent bundle**, and is denoted TU. A (smooth, contravariant) **vector field** on U is then a tangent vector defined at each point such that its application to a smooth function on U is again smooth. Similarly, a **covariant vector field** is a 1-form defined at each point such that its value on a vector field is a smooth function, and a **tensor field** is the tensor product of vector fields and covariant vector fields.

△ Tensor fields (including vector fields and covariant vector fields) are written using the same notation as tensors, making it important to distinguish the two situations. In particular, one can define a (pseudo) metric tensor field, which is then usually referred to as simply a metric.

Note that a tensor field must remain a tensor locally at any point p, i.e. it must be a multi-linear mapping. For example, a covariant tensor field can only depend upon the values of its vector field arguments at p, since otherwise one could add a vector field that vanishes at p and obtain a different result. This means that operators such as the derivatives on manifolds we will see in Section 6.3 and Chapter 9 cannot usually be viewed as tensors, since they measure the difference between arguments at different points.

Since vectors are operators on functions, we can apply one vector field to another. Following the practice of using $\partial/\partial x^u$ to refer to $\partial/\partial a^\mu$, this can be used to define the **Lie bracket of vector fields**

$$[v, w](f) \equiv v(w(f)) - w(v(f))$$

$$\Rightarrow [v, w] = \left(v^\mu \frac{\partial w^\nu}{\partial x^\mu} - w^\mu \frac{\partial v^\nu}{\partial x^\mu} \right) \frac{\partial}{\partial x^\nu}.$$

Here we have used the equality of mixed partials, and can easily verify that $[v, w]$ is anti-commuting and satisfies the Jacobi identity. Since this expression is coordinate-independent, $[v, w]$ is a vector field and we can thus view vect (M), the set of all vector fields on M, as the infinite-dimensional **Lie algebra of vector fields** on M, with vector multiplication defined by the Lie bracket.

Having defined vector and tensor fields on manifolds, we can now define a **differential form** as an alternating covariant tensor field, i.e. an exterior form in $\Lambda(T_pU)$ smoothly defined for every point p.

$\triangle$ Just as tensor fields are usually referred to as simply tensors, differential forms are usually referred to as simply **forms**, and a k-form is written simply $\varphi \in \Lambda^k M$. It is important to remember that in the context of manifolds, a k-form is an exterior form smoothly defined on k elements of the tangent space at each point, i.e. an anti-symmetric covariant k-tensor field.

On a differentiable manifold, the existence of k-forms makes possible a more concrete definition of orientability: a manifold M^n is orientable iff there exists a non-vanishing n-form. Such a form is called a **volume form** (AKA volume element), since as we recall from Section 3.1.3 it gains a Jacobian-like determinant factor under invertible linear transformations.

$\triangle$ The term "volume form" or "volume element" is sometimes defined in physics to reflect the intuitive idea of a form which returns the volume spanned by its argument vectors; however, volume is always positive, so that in this usage we are more accurately referring to a **volume pseudo-form** whose value is the absolute value of the volume form as we have defined it.

6.1.3 Frames

A **frame** e_μ on $U \subset M^n$ is defined to be a tensor field of bases for the tangent spaces at each point, i.e. n linearly independent smooth vector fields e_μ.

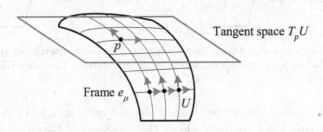

Tangent space $T_p U$

Frame e_μ

U

Figure 6.1.4 A frame e_μ is n smooth vector fields that together provide a basis for the tangent space at every point.

The concept of frame has a particularly large number of synonyms, including comoving frame, repère mobile, vielbein, n-frame, and n-bein (where n is the dimension). The **dual frame**, the 1-forms β^μ corresponding to a frame e_μ, is also often simply called the frame.

When using particular coordinates x^μ, the frame $e_\mu = \partial/\partial x^\mu$ is called the **coordinate frame** (AKA coordinate basis or associated basis); any other frame is then called a **non-coordinate frame**. A **holonomic frame** is a coordinate frame in some coordinates (though perhaps not the ones being used); this condition is equivalent to requiring that $[e_\mu, e_\nu] = 0$, a result which is sometimes called **Frobenius' theorem**. An **anholonomic frame** is then a frame that cannot be derived from any coordinate chart in its region of definition. Using a non-coordinate frame suited to a specific problem is sometimes called the **method of moving frames**.

$\triangle$ Note that the distinction between holonomic and coordinate frames as defined here is often not made.

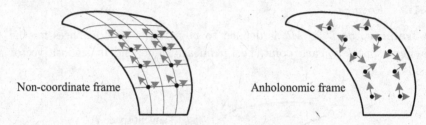

Figure 6.1.5 A non-coordinate frame is not tangent to the coordinate functions being used, while an anholonomic frame cannot be derived from any coordinate chart.

A frame cannot usually be globally defined on a manifold. A simple way to see this is by the example of the 2-sphere S^2. Any drawing of coordinate functions on a globe will have singularities, such as the north and south poles when using latitude and longitude; these are points where the associated coordinate frame will either be undefined or will vanish. In general, there is no non-zero smooth vector field that can be defined on S^n for even n (this is sometimes called the **hedgehog theorem**, AKA hairy ball theorem, coconut theorem).

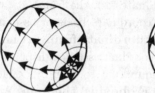

Figure 6.1.6 The hedgehog theorem for S^2, showing that any attempt to "comb the hair of a hedgehog" yields bald spots, in this case at the poles.

A manifold that can have a global frame defined on it is called **parallelizable**. Some facts regarding parallelizable manifolds include:

- All parallelizable manifolds are orientable (and therefore have a volume form), but as we saw with S^2 the converse is not in general true
- Any orientable 3-manifold M^3 is parallelizable $\Rightarrow$ any 4-manifold $M^3 \times \mathbb{R}$ is parallelizable (important in the case of the spacetime manifold)
- Of the n-spheres, only S^1, S^3, and S^7 are parallelizable (this can be seen to be related to $\mathbb{C}$, $\mathbb{H}$, and $\mathbb{O}$ being the only normed finite-dimensional real division algebras beyond $\mathbb{R}$)

- The torus is the only closed orientable surface with a non-zero smooth vector field

6.1.4 *Tangent vectors in terms of frames*

It is important to remember that in following our intuitive picture of a Euclidean surface, our central definitions were manifolds M and tangent vectors v. These are the "real" intrinsic objects, while their expressions in terms of a particular coordinate chart and frame are arbitrary. Coordinates and frames are "temporary" tools we use to "componentize" points and tangents on a manifold.

In particular, if a manifold is defined in terms of a set of coordinate functions that feature a singularity, this singularity may be due to the coordinates extending outside of their valid chart, telling us nothing about whether the manifold itself has a singularity. Every point of a well-defined differentiable manifold always has a local coordinate chart and tangent vectors.

For example, given the typical spherical coordinate chart for S^2 the associated frame will be singular at the poles, since they are outside of U for that chart; nevertheless, tangent vectors are well-defined at these points, and can be expressed perfectly normally in a different chart.

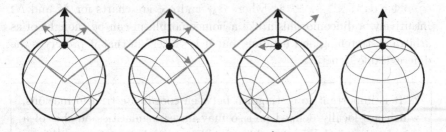

Figure 6.1.7 A manifold and tangent vector expressed in terms of different coordinate functions and frames.

In the above figure, we see the following situations depicted:

- $v = e_1 + e_2 = \partial/\partial x^1 + \partial/\partial x^2$ (expressed in a coordinate frame)
- $v = e_1' - e_2' = \partial/\partial x'^1 - \partial/\partial x'^2$ (using a different coordinate frame)
- $v = e_1'' + 3e_2'' = \partial/\partial x'^1 - \partial/\partial x'^2$ (in a non-coordinate frame)

The final figure depicts coordinate functions that are singular at the point of interest; the manifold and vector are still well-defined, but the tangent space at this point cannot be expressed in terms of this coordinate chart.

$\triangle$ In general, when working with objects on manifolds, it is important to keep clearly in mind whether a given symbol represents a vector, form, or function (0-form); whether any given index is a label, an abstract index or a component index in a particular frame or coordinates; and whether the object is a field with a value at each point, or is only valid at a particular point. Any calculation can always be made explicit by expressing everything in terms of functions and differential operators on them.

6.2 Mapping manifolds

6.2.1 *Diffeomorphisms*

In the same way that spaces or topological manifolds are equivalent if they are related by a homeomorphism, differentiable manifolds are equivalent if they are related by a **diffeomorphism**, a homeomorphism that is differentiable along with its inverse. As usual we define differentiability by moving the mapping to $\mathbb{R}^n$, e.g. $\Phi\colon M \to N$ is differentiable if $\alpha_N \circ \Phi \circ \alpha_M^{-1}\colon \mathbb{R}^m \to \mathbb{R}^n$ is, where α_M and α_N are charts for M and N. Intuitively, a diffeomorphism like a homeomorphism can be thought of as arbitrary stretching and bending, but it is "nicer" in that it preserves the differentiable structure.

$\triangle$ It is important to distinguish between coordinate transformations, which are locally defined and so may have singularities outside of a given region; and diffeomorphisms, which are globally defined and form a group. One can define a coordinate transformation on a region of a manifold that avoids any resulting singularities, but a diffeomorphism must be smooth on the entire manifold.

6.2.2 *The differential and pullback*

If we consider a general mapping between manifolds $\Phi\colon M^m \to N^n$, we can choose charts $\alpha_M\colon M \to \mathbb{R}^m$ and $\alpha_N\colon N \to \mathbb{R}^n$, with coordinate functions x^μ and y^ν, so that the mapping $\alpha_N \circ \Phi\colon M \to \mathbb{R}^n$ can be represented by n

functions $\Phi^\nu \colon M \to \mathbb{R}$. This allows us to write down an expression for the induced **tangent mapping** or **differential** (aka pushforward, derivative) $d\Phi \colon TM \to TN$ (also denoted $T\Phi$ or Φ_* or sometimes simply Φ if it is clear the argument is a tangent vector). For a tangent vector $v = v^\mu \partial/\partial x^\mu$ at a point $p \in M$ we define

$$d\Phi\left(v\right)\big|_p \equiv v^\mu \frac{\partial \Phi^\nu}{\partial x^\mu} \frac{\partial}{\partial y^\nu}\bigg|_{\Phi(p)}.$$

This definition can be shown to be coordinate-independent and to follow our intuitive expectation that mapped tangent vectors stay tangent to mapped curves. If $M = N$ and Φ is the identity, $d\Phi$ is just the vector component transformation in Section 6.1.2. The matrix $J_\Phi(x) \equiv \partial \Phi^\nu/\partial x^\mu$ is called the **Jacobian matrix** (AKA Jacobian).

If Φ is a diffeomorphism, $d\Phi$ is an isomorphism between the tangent spaces at every point in M. The **inverse function theorem** says that the converse is true locally: if $d\Phi_p$ is an isomorphism at $p \in M$, then Φ is locally a diffeomorphism. In particular, this means that if in some coordinates the Jacobian matrix is nonsingular, then $\alpha_N \circ \Phi \circ \alpha_M^{-1}$ represents a locally valid coordinate transformation and $\Phi^\nu = y^\nu$.

A mapping between manifolds $\Phi \colon M^m \to N^n$ also can be used to naturally define the **pullback** of a form $\Phi^* \colon \Lambda^k N \to \Lambda^k M$ by $\Phi^*\varphi\left(v_1, \ldots, v_k\right) = \varphi(d\Phi\left(v_1\right), \ldots, d\Phi\left(v_k\right))$, where the name indicates that a form on N can be "pulled back" to M using Φ. Note that the composition of pullbacks is then $\Psi^*\Phi^*\varphi = (\Phi\Psi)^*\varphi$.

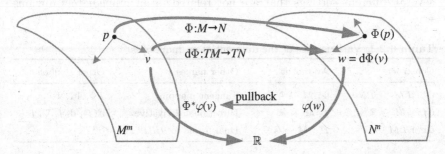

Figure 6.2.1 Forms φ on N are pulled back to M by sending argument vectors to N using $d\Phi$.

Note that for a mapping $f \colon M \to \mathbb{R}$, we have $df \colon TM \to T\mathbb{R} \cong \mathbb{R}$, so that $df\left(v\right) = v^\mu \partial f/\partial x^\mu = v\left(f\right)$, the directional derivative of f. Let us apply this to the coordinate function $x^1 \colon M \to \mathbb{R}$. Then we have $\mathrm{d}x^1\,(v) = v^\mu \partial x^1/\partial x^\mu = v^1$, so that in particular $\mathrm{d}x^\nu\,(\partial/\partial x^\mu) = \delta^\nu{}_\mu$, i.e. $\mathrm{d}x^\mu$ is in fact the dual frame to $\partial/\partial x^\mu$. Thus in a given coordinate system, we can write a general tensor of type (m, n) as

$$T = T^{\mu_1\cdots\mu_m}{}_{\nu_1\cdots\nu_n} \frac{\partial}{\partial x^{\mu_1}} \otimes \cdots \otimes \frac{\partial}{\partial x^{\mu_m}} \otimes \mathrm{d}x^{\nu_1} \otimes \cdots \otimes \mathrm{d}x^{\nu_n}.$$

In particular, the metric tensor is often written $\mathrm{d}s^2 \equiv g = g_{\mu\nu}\mathrm{d}x^\mu \mathrm{d}x^\nu$, where the Einstein summation convention is used and the tensor symbol omitted. A general k-form $\varphi \in \Lambda^k M$ can then be written as

$$\varphi = \sum_{\mu_1 < \cdots < \mu_k} \varphi_{\mu_1\ldots\mu_k} \mathrm{d}x^{\mu_1} \wedge \cdots \wedge \mathrm{d}x^{\mu_k}.$$

From either the tangent mapping definition or the behavior of the exterior product under a change of basis from Section 3.1.3, we see that under a change of coordinates we have

$$\mathrm{d}y^{\mu_1} \wedge \cdots \wedge \mathrm{d}y^{\mu_k} = \det\left(\frac{\partial y^\nu}{\partial x^\mu}\right) \mathrm{d}x^{\mu_1} \wedge \cdots \wedge \mathrm{d}x^{\mu_k}.$$

This is the familiar **Jacobian determinant** (like the Jacobian matrix, also often called the Jacobian) that appears in the change of coordinates rule for integrals from calculus, and explains the name of the volume form as defined previously in terms of the exterior product.

In summary, the differential d has a single definition, but is used in several different settings that are not related in an immediately obvious way.

Table 6.2.1 Various uses of the differential on manifolds.

Construct	Argument	Other names	Other symbols
$\mathrm{d}\Phi\colon TM \to TN$	$\Phi\colon M \to N$	Tangent mapping	$T\Phi,\ \Phi_*,\ \Phi$
$\mathrm{d}f\colon TM \to \mathbb{R}$	$f\colon M \to \mathbb{R}$	Directional derivative	$v\,(f),\ \mathrm{d}_v f,\ \nabla_v f$
$\mathrm{d}x^\mu\colon TM \to \mathbb{R}$	$x^\mu\colon M \to \mathbb{R}$	Dual frame to $\partial/\partial x^\mu$	β^μ

6.2.3 *Immersions and embeddings*

We can generalize and make precise the concept of a surface embedded in 3-dimensional space with the following definitions concerning a differentiable map $\Phi\colon M^m \to N^n$:

- **Immersion**: $d\Phi$ is injective for all $p \in M$; intuitively, a smooth mapping that doesn't collapse the tangent spaces
- **Submanifold**: an immersion with Φ injective; intuitively, an immersion that doesn't intersect itself
- **Embedding** (AKA imbedding): a submanifold with Φ a homeomorphism onto $\Phi(M)$; intuitively, a submanifold that doesn't have intersecting limit points

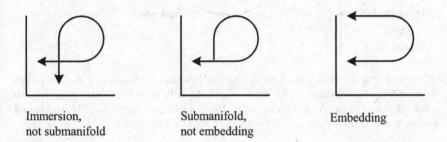

Immersion,   Submanifold,   Embedding
not submanifold   not embedding

Figure 6.2.2 $\mathbb{R}$ immersed in $\mathbb{R}^2$; the second immersion approaches a self-intersection in the limit as the line approaches infinity.

The difference in dimension $(n - m)$ is called the **codimension** of the embedding. The **Whitney embedding theorem** states that any M^m can be immersed in $\mathbb{R}^{(2m-1)}$ and embedded in $\mathbb{R}^{2m}$. Thus we can view differentiable manifolds as generalized surfaces that we study without making reference to the enclosing Euclidean space. The limiting dimension of this theorem is illustrated by noting that the real projective space $\mathbb{R}P^m$ cannot be embedded in $\mathbb{R}^{(2m-1)}$.

6.2.4 Critical points

General mappings Φ between manifolds classify points according to how they transform, and can be used to extract information about the manifolds:

- **Regular point**: $p \in M$ such that $d\Phi_p$ maps T_pM onto T_pN; if the map is not onto, p is called a **critical point**
- **Regular value**: $q \in N$ such that $\Phi^{-1}(q)$ consists of all regular points or is empty; if $\Phi^{-1}(q)$ includes a critical point, q is called a **critical value**

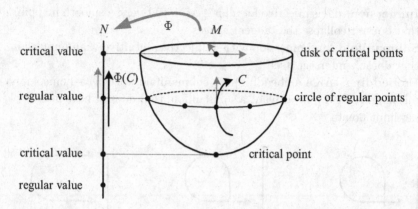

Figure 6.2.3 Critical points of the height function Φ mapping a hollow bullet to its vertical component. The flat top and tip only have horizontal tangents, so that dΦ is not onto. At a regular point, the tangent to a curve $C \in M$ is mapped to the tangent of the mapped curve $\Phi(C)$ via the Jacobian.

Sard's theorem states that if Φ is sufficiently differentiable, almost all values are regular (we will not elaborate on "sufficient" and "almost" here). **Morse theory** uses these concepts to extract cell complex structures and homological information from a given manifold.

6.3 Derivatives on manifolds

In this section we will introduce various objects that in some way measure how vectors or forms change from point to point on a manifold.

6.3.1 *Derivations*

In general, we define a **derivation** to be a linear map $\mathcal{D}\colon \mathfrak{a} \to \mathfrak{a}$ on an algebra $\mathfrak{a}$ that follows the **Leibniz rule** (AKA product rule)

$$\mathcal{D}(AB) = (\mathcal{D}A)B + A(\mathcal{D}B).$$

As noted previously in Section 6.1.2, the set vect(M) of vector fields on a manifold form a Lie algebra; the Lie bracket operation with a fixed vector field $[u, \ ]$ is then a derivation on this algebra, since the Leibniz rule

$$[u, [v, w]] = [[u, v], w] + [v, [u, w]]$$

is just the Jacobi identity.

For a graded algebra, e.g. the exterior algebra, the **degree** of a derivation is the integer c where $\mathcal{D}\colon \Lambda^k M \to \Lambda^{k+c} M$. A **graded derivation** is defined to follow the **graded Leibniz rule**, e.g. for a k-form φ,

$$\mathcal{D}\left(\varphi \wedge \psi\right) = \mathcal{D}\varphi \wedge \psi + (-1)^{kc}\,\varphi \wedge \mathcal{D}\psi.$$

If c is odd, a graded derivation is sometimes called an **anti-derivation** (AKA skew-derivation).

6.3.2 The Lie derivative of a vector field

Without some kind of additional structure, there is no way to "transport" vectors, or compare them at different points on a manifold, and therefore no way to construct a vector derivative. The simplest way to introduce this structure is via another vector field, which leads us to the **Lie derivative** $L_v w \equiv [v, w]$; as noted above, L_v is a derivation due to the Jacobi identity. In this section we define the Lie derivative in terms of infinitesimal vector transport, and explore its geometrical meaning.

Given any vector field v on M^n, it can be shown (Frankel [1997] pp. 125-127) that there exists a parameterized curve $v_p(t)$ at every point $p \in M$ such that $v_p(0) = p$ and $\dot{v}_p(t)$ is the value of the vector field v at the point $v_p(t)$ (the dot indicates the derivative with respect to t, which as usual is calculated on the curve mapped to $\mathbb{R}^n$ by the coordinate chart). Each curve in this family is in general only well-defined locally, i.e. for $-\varepsilon < t < \varepsilon$, and is thus called the **local flow** of v.

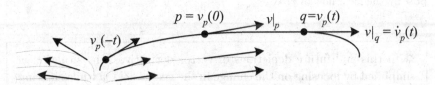

Figure 6.3.1 A depiction of the local flow of a vector field v, with details on the local parameterized curve $v_p(t)$ at a point p.

For a fixed value of t, there is some region $U \subset M$ where the map $\Phi_t\colon U \to U$ defined by $p \mapsto v_p(t)$ is a diffeomorphism, and within the valid domain of t the maps Φ_t satisfy the abelian group law $\Phi_t \circ \Phi_s = \Phi_{t+s}$; thus the Φ_t are called a **local one-parameter group of diffeomorphisms**.

This name is somewhat misleading, since due to the limited domain of t the maps Φ_t do not actually form a group; the "local" reflects the fact that the diffeomorphisms are not on all of M. In the case that these maps are in fact valid for all of t and M, v is called a **complete vector field**, and the Φ_t are called a **one-parameter group of diffeomorphisms**. If M is compact, then every vector field is complete; if not, then a vector field is complete if it has **compact support** (is zero except on a compact subset of M).

The tangent map $\mathrm{d}\Phi$ defined by the vector field v is then the extra structure we need to "transport" vectors. $\mathrm{d}\Phi$ maps a vector tangent to the curve C to a vector tangent to the curve $\Phi(C)$; it "pushes vectors along the flow of v." We can now define the Lie derivative as a limit

$$\dot{L}_v w \equiv \lim_{\varepsilon \to 0} \frac{1}{\varepsilon} \left[\mathrm{d}\Phi_{-\varepsilon} \left(w \big|_{v_p(\varepsilon)} \right) - w \big|_p \right]$$

$$= \lim_{\varepsilon \to 0} \frac{1}{\varepsilon} \left[w \big|_{v_p(\varepsilon)} - \mathrm{d}\Phi_\varepsilon \left(w \big|_p \right) \right].$$

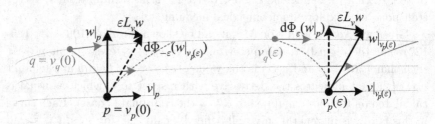

Figure 6.3.2 The Lie derivative $L_v w$ is "the difference between w and its transport by the local flow of v."

> ✷ In this and future depictions of vector derivatives, the situation is simplified by focusing on the change in the vector field w while showing the "transport" of w as a parallel displacement. This has the advantage of highlighting the equivalence of defining the derivative at either 0 or ε in the limit $\varepsilon \to 0$. Depicting $L_v w$ as a non-parallel vector at $v_p(t)$ would be more accurate, but would obscure this fact. We also will follow the picture here in using words to characterize derivatives: namely, "the difference" is short for "the difference per unit ε to order ε in the limit $\varepsilon \to 0$."

This definition can be shown to be equivalent to $L_v w \equiv [v, w]$. Another way of depicting the Lie derivative that highlights the anti-commutativity of the Lie bracket is to consider $L_v w$ in terms of a loop defined by the flows of v and w.

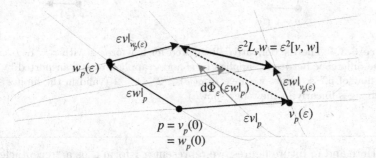

Figure 6.3.3 The Lie derivative $L_v w$ can also be pictured as the vector field whose local flow is the "commutator of the flows of v and w," i.e. it is the difference between the local flow of v followed by w and that of w followed by v. Thus $L_v w$ "completes the parallelogram" formed by the flow lines.

6.3.3   The Lie derivative of an exterior form

The Lie derivative L_v can be applied to a k-form φ by using the pullback of φ by the diffeomorphism Φ associated with the flow of v, i.e. applied to k vectors w_I we define

$$L_v \varphi\,(w_I) \equiv \lim_{\varepsilon \to 0} \frac{1}{\varepsilon} \left[\varphi\,(\mathrm{d}\Phi_\varepsilon\,(w_I)) - \varphi\,(w_I) \right].$$

The Lie derivative is thus a derivation of degree 0 on the exterior algebra. $L_v \varphi$ measures the change in φ as its arguments are transported by the local flow of v. In the case of a 0-form f, this is just the differential or directional derivative $L_v f = v(f) = \mathrm{d}f(v)$.

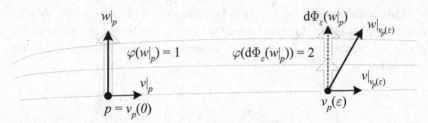

Figure 6.3.4 The Lie derivative illustrated for a 1-form φ with $\varepsilon = 1$. $L_v\varphi(w)$ is "the difference between φ applied to w and φ applied to w transported by the local flow of v," so above we have $L_v\varphi(w) = 2 - 1 = 1$ (valid in the limit $\varepsilon \to 0$ if φ changes linearly in the range shown).

✿ Here and in future figures, we represent a 1-form φ as a "receptacle" $\varphi^\Uparrow \equiv \varphi^\sharp / \left\| \varphi^\sharp \right\|^2$ which when applied to a vector "arrow" argument v yields the number of receptacles covered by the projection of v onto $\varphi^\sharp$, which is the value of $\varphi(v)$. This can be seen by recalling from Section 3.2.1 that $\varphi(v)/\left\|\varphi^\sharp\right\|$ is the length of the projection of v onto $\varphi^\sharp$, so that this projection divided by the length of the receptacle $\left\|\varphi^\Uparrow\right\| = 1/\left\|\varphi^\sharp\right\|$ recovers the value $\varphi(v)$. The advantage of this approach is that values can be calculated from the figure absent a length scale. Another common graphical device is to represent 1-forms as "surfaces" which are "pierced" by the arrows.

△ The common practice of depicting a 1-form φ in terms of the associated vector $\varphi^\Uparrow$ as above has consequences that can be non-intuitive. For example, doubling the value of the 1-form means halving its length in the illustration, i.e. the value of the 1-form can be viewed as the "density" of receptacles. Also, when depicting φ as changing linearly, the length L of the 1-form representation changes like $L \mapsto L/(1 + \varepsilon)$, which doesn't appear linear as a vector representation would, whose length changes like $L \mapsto L(1 + \varepsilon)$.

By using the above definitions of the Lie derivative applied to vectors and 1-forms, we can extend it to tensors by using these definitions for each component. In a holonomic frame, we can obtain an expression for the Lie derivative of a tensor in terms of coordinates

$$L_w T^{\mu_1 \ldots \mu_m}{}_{\nu_1 \ldots \nu_n} = w^\lambda \frac{\partial}{\partial x^\lambda} T^{\mu_1 \ldots \mu_m}{}_{\nu_1 \ldots \nu_n}$$

$$- \sum_{j=1}^{m} \left(\frac{\partial}{\partial x^\lambda} w^{\mu_j} \right) T^{\mu_1 \ldots \mu_{j-1} \lambda \mu_{j+1} \ldots \mu_m}{}_{\nu_1 \ldots \nu_n}$$

$$+ \sum_{j=1}^{n} \left(\frac{\partial}{\partial x^{\nu_j}} w^\lambda \right) T^{\mu_1 \ldots \mu_m}{}_{\nu_1 \ldots \nu_{j-1} \lambda \nu_{j+1} \ldots \nu_n}.$$

6.3.4 The exterior derivative of a 1-form

The Lie derivative $L_v \varphi$ is defined in terms of a vector field v, and its value as a "change in φ" is computed by using v to transport the arguments of φ. In contrast, recall that the differential d takes a 0-form $f \colon M \to \mathbb{R}$ to a 1-form $df \colon TM \to \mathbb{R}$ with $df(v) = v(f)$. Thus d is a derivation of degree $+1$ on 0-forms, whose value as a "change in f" is computed using the vector field argument of the resulting 1-form.

We would like to generalize d to k-forms by extending this idea of including the "direction argument" by increasing the degree of the form. It turns out that if we also require the property $d\left(d\left(\varphi\right)\right) = 0$ (or "$d^2 = 0$"), there is a unique graded derivation of degree $+1$ that extends d to general k-forms; this derivation is called the **exterior derivative**. We first explore the exterior derivative of a 1-form.

The exterior derivative of a 1-form is defined by

$$d\varphi\left(v, w\right) \equiv v\left(\varphi\left(w\right)\right) - w\left(\varphi\left(v\right)\right) - \varphi\left([v, w]\right),$$

where e.g.

$$v\left(\varphi\left(w\right)\right) = \lim_{\varepsilon \to 0} \frac{1}{\varepsilon} \left[\varphi\left(w \big|_{v_p(\varepsilon)}\right) - \varphi\left(w \big|_p\right)\right]$$

measures the change of $\varphi\left(w\right)$ in the direction v, so that

$$d\varphi\left(v, w\right) = \lim_{\varepsilon \to 0} \frac{1}{\varepsilon^2} \left[\left(\varphi\left(\varepsilon w \big|_{v_p(\varepsilon)}\right) - \varphi\left(\varepsilon w \big|_p\right)\right)\right.$$

$$- \left(\varphi\left(\varepsilon v \big|_{w_p(\varepsilon)}\right) - \varphi\left(\varepsilon v \big|_p\right)\right)$$

$$\left. - \varphi\left(\varepsilon^2 \left[v, w\right]\right)\right].$$

The term involving the Lie bracket "completes the parallelogram" formed by v and w, so that $d\varphi\left(v, w\right)$ can be viewed as the "sum of φ on the boundary of the surface defined by its arguments."

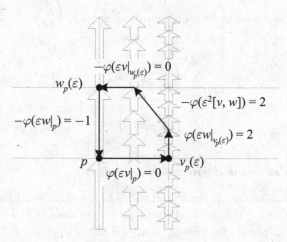

Figure 6.3.5 The exterior derivative of a 1-form $d\varphi\,(v,w)$ is the sum of φ along the boundary of the completed parallelogram defined by v and w. So if in the diagram $\varepsilon = 1$, we have $d\varphi\,(v,w) = (2-1) - (0-0) + 2 = 3$. This value is valid in the limit $\varepsilon \to 0$ if the sum varies like ε^2 as depicted in the figure.

The identity $d^2 = 0$ can then be seen as stating the intuitive fact that the boundary of a boundary is zero. If $\varphi = df$, then $\varphi\,(v) = df\,(v) = v\,(f)$, the change in f along v. Thus e.g. $\varepsilon\varphi\,(v\,|_p) = f\,(v_p\,(\varepsilon)) - f\,(p)$, so that the value of φ on v is the difference between the values of f on the two points which are the boundary of v. Each endpoint will be cancelled by a starting point as we add up values of φ along a sequence of vectors, resulting in the difference between the values of f at the boundary of the total path defined by these vectors. $d\varphi$ is the value of φ over the boundary path of the surface defined by its arguments, which has no boundary points and so vanishes.

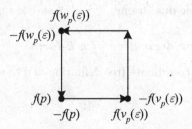

Figure 6.3.6 $d^2 = 0$ corresponds to the boundary of a boundary is zero: each term $\varphi(v) = df(v)$ is the difference between the values of f on the boundary points of v, which cancel as we traverse the boundary of the surface defined by the arguments of $d\varphi(v,w)$. In the figure we assume a vanishing Lie bracket for simplicity.

Note that $d\varphi(v,w)$ measures the interaction between φ and the vector fields v and w, thus avoiding the need to "transport" vectors. In particular, a non-zero exterior derivative can be pictured as resulting from either the vector fields or φ "changing," i.e. changing with regard to the implied coordinates of our pictures.

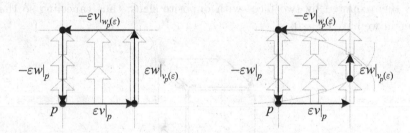

Figure 6.3.7 A non-zero exterior derivative $d\varphi(v,w)$ results from changes in $\varphi(v)$ or $\varphi(w)$, not changes in either φ or the vector fields alone as compared to some transport.

If we calculate $d\varphi(e_1, e_2)$ explicitly in a holonomic frame in two dimensions, $d\left(\varphi_1 dx^1 + \varphi_2 dx^2\right) = d\varphi_1 \wedge dx^1 + d\varphi_2 \wedge dx^2$, so applying this to the basis vector fields e_1 and e_2 we have

$$
\begin{aligned}
d\varphi(e_1, e_2) &= d\varphi_1(e_1) \cdot dx^1(e_2) - d\varphi_1(e_2) \cdot dx^1(e_1) \\
&\quad + d\varphi_2(e_1) \cdot dx^2(e_2) - d\varphi_2(e_2) \cdot dx^2(e_1) \\
&= e_1(\varphi_2) - e_2(\varphi_1) \\
&= \partial/\partial x^1(\varphi_2) - \partial/\partial x^2(\varphi_1).
\end{aligned}
$$

Note that a holonomic dual frame $\beta^\mu = \mathrm{d}x^\mu$ satisfies $\mathrm{d}\beta^\mu = \mathrm{dd}x^\mu = 0$.

6.3.5   *The exterior derivative of a k-form*

The extension of the coordinate-free definition of d to general k-forms gives the expression

$$
\begin{aligned}
\mathrm{d}\varphi\,(v_0, \ldots, v_k) \\
\equiv \sum_{j=0}^{k} (-1)^j\, v_j\,(\varphi\,(v_0, \ldots, v_{j-1}, v_{j+1}, \ldots, v_k)) \\
+ \sum_{i<j} (-1)^{i+j}\, \varphi\,([v_i, v_j]\,, v_0, \ldots, v_{i-1}, v_{i+1}, \ldots, v_{j-1}, v_{j+1}, \ldots, v_k)\,.
\end{aligned}
$$

Our picture for 1-forms then extends to higher dimensions using much the same reasoning as used for homology in Section 5.2. For example, assuming vanishing Lie brackets to simplify the picture, the exterior derivative of a 2-form $\mathrm{d}\varphi\,(u, v, w)$ can be viewed as the "sum of φ on the boundary faces of the cube defined by its arguments." If $\varphi = \mathrm{d}\psi\,(v, w)$ is the boundary of a face, $\mathrm{d}\varphi = \mathrm{d}^2\psi$ is the sum of the boundaries of the faces; each edge is then counted by two faces with opposite signs, thus canceling so that again we have $\mathrm{d}^2 = 0$.

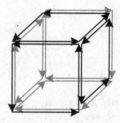

Figure 6.3.8 The 3-form $\mathrm{d}\varphi = \mathrm{d}^2\psi$ sums ψ over the edges of the faces of a cube. The sum vanishes since each edge is counted twice with opposite signs.

The similarity between the exterior derivative d and the boundary homomorphism ∂ from homology is no illusion, as we will see shortly in Section 6.4.

In a holonomic frame, we can obtain an expression for $\mathrm{d}\varphi$ in terms of coordinates

$$\mathrm{d}\varphi = \sum_{\mu_0 < \cdots < \mu_k} \left(\sum_{j=0}^{k} (-1)^j \frac{\partial}{\partial x^{\mu_j}} \varphi_{\mu_0 \ldots \mu_{j-1}\mu_{j+1} \ldots \mu_k} \right) \mathrm{d}x^{\mu_0} \wedge \cdots \wedge \mathrm{d}x^{\mu_k},$$

or even more explicitly,

$$\mathrm{d}\varphi = \sum_{\mu_0 < \cdots < \mu_k} \left(\frac{\partial}{\partial x^{\mu_0}} \varphi_{\mu_1 \ldots \mu_k} - \frac{\partial}{\partial x^{\mu_1}} \varphi_{\mu_0 \mu_2 \ldots \mu_k} + \cdots \right.$$
$$\left. + (-1)^k \frac{\partial}{\partial x^{\mu_k}} \varphi_{\mu_0 \ldots \mu_{k-1}} \right) \mathrm{d}x^{\mu_0} \wedge \cdots \wedge \mathrm{d}x^{\mu_k}.$$

It is not hard to see that the exterior derivative commutes with the pullback, i.e. $\Phi^* \mathrm{d}\varphi = \mathrm{d}\Phi^* \varphi$.

$\triangle$ Despite a convenient description using coordinates associated with a holonomic frame, it is important to keep in mind that the exterior derivative of a form is frame- and coordinate-independent.

If we include an inner product, vector calculus can be seen to correspond to exterior calculus on $\mathbb{R}^3$, and can thus be generalized to arbitrary dimensions:

- For a function (0-form) f, the components of the 1-form $\mathrm{d}f$ correspond to those of the gradient of f, i.e. $(\mathrm{d}f)_\mu = (\nabla f)^\mu$ or $\nabla f = (\mathrm{d}f)^\sharp$; a generalization of the gradient is then the 1-form $\mathrm{d}f$
- For a 1-form with components equal to those of a vector $\varphi_\mu = v^\mu$, the components of $\mathrm{d}\varphi$ correspond to those of the curl of v, i.e. $(\mathrm{d}\varphi)_\mu = (\nabla \times v)^\mu$ or $(\nabla \times v) = (*\mathrm{d}(v^\flat))^\sharp$; a generalization of the curl is then the 2-form $\mathrm{d}\varphi$
- For a 2-form with components equal to those of a vector $\psi_\mu = (*\varphi)_\mu = v^\mu$, the value of $\mathrm{d}\psi$ corresponds to the value of the divergence of v, i.e. $\mathrm{d}\psi = \nabla \cdot v$ or $\nabla \cdot v = *\mathrm{d}(*(v^\flat))$; a generalization of the divergence is then the value $*\mathrm{d}(*\varphi)$

In $\mathbb{R}^3$ the relations curl grad = div curl = 0 thus correspond to the property $\mathrm{d}^2 = 0$. Note that we have used the musical isomorphisms on $\mathbb{R}^3$, which imply an inner product, as does the Hodge star.

Finally, the classical gradient, curl, and divergence integral theorems in vector calculus are generalized to **Stokes' theorem**: for an $(n-1)$-form φ on a compact oriented manifold M^n with boundary ∂M,

$$\int_M \mathrm{d}\varphi = \int_{\partial M} \varphi.$$

This is essentially the integral form of the property $\mathrm{d}^2 = 0$: summing $\mathrm{d}\varphi$ over M can be pictured as summing φ over the boundaries of infinitesimal volumes, so that all internal boundaries cancel and what is left is φ over the outer boundary ∂M.

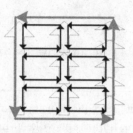

Figure 6.3.9 The integral of $\mathrm{d}\varphi$ over M can be pictured as summing φ over the boundaries of infinitesimal volumes, so that all internal boundaries cancel and what is left is φ over the outer boundary ∂M.

We will not address the details of defining integration on manifolds here, but the basic idea is relatively straightforward: a coordinate chart maps an n-dimensional sub-manifold of M^n to $S \in \mathbb{R}^n$; an n-form φ can then be written $f(x_I)\,\mathrm{d}x^I$, and its integral is defined to be $\int_S f(x_I)\,\mathrm{d}x^I$, which can be shown to be coordinate-independent. Note that without additional structure on the manifold, we cannot integrate functions or other forms over M^n besides n-forms.

6.3.6   *Relationships between derivations*

We can define one other derivation on k-forms, the **interior derivative** (AKA inner derivative, inner multiplication), which is the generalization of the interior product to forms on manifolds; i.e. for a given vector v it is the graded degree -1 derivation $(i_v\varphi)(w_2,\ldots,w_k) \equiv \varphi(v,w_2,\ldots,w_k)$ on k-forms φ, which follows the graded Leibniz rule $i_v(\varphi \wedge \psi) = (i_v\varphi) \wedge \psi + (-1)^k \varphi \wedge (i_v\psi)$. The graded commutativity of forms immediately gives the property $i_v i_w + i_w i_v = i_v^2 = 0$.

The interior, exterior, and Lie derivatives then form an infinite-dimensional graded Lie algebra with the following relations:

- $[L_v, L_w] \equiv L_v L_w - L_w L_v = L_{[v,w]}$
- $[i_v, i_w] \equiv i_v i_w + i_w i_v = 0$
- $[\mathrm{d}, \mathrm{d}] \equiv \mathrm{d}^2 + \mathrm{d}^2 = 0$
- $[L_v, i_w] \equiv L_v i_w - i_w L_v = i_{[v,w]}$
- $[L_v, \mathrm{d}] \equiv L_v \mathrm{d} - \mathrm{d} L_v = 0$
- $[i_v, \mathrm{d}] \equiv i_v \mathrm{d} + \mathrm{d} i_v = L_v$

6.4 Homology on manifolds

The additional structure of coordinates and tangents can be used to revisit homology, gaining additional insight and results. In particular, as we saw in Section 6.3.5, the exterior derivative d exhibits structure reminiscent of the boundary homomorphism ∂ in homology. This can be exploited to build a version of homology based on forms instead of on simplices.

6.4.1 *The Poincaré lemma*

We first define the form versions of cycles and boundaries in homology:

- **Closed form:** an element of Ker d, i.e. a form φ such that $\mathrm{d}\varphi = 0$
- **Exact form:** an element of Im d, i.e. a form φ such that $\varphi = \mathrm{d}\psi$

In this context, the term **Poincaré lemma** can refer to either the property "exact $\Rightarrow$ closed" ($\mathrm{d}^2 = 0$), or the converse statement "closed $\Rightarrow$ exact" under certain topological conditions, which we address next.

Let us try to picture a closed 1-form that is not exact using a coordinate frame. If the 1-form φ is closed, it integrates to zero around any coordinate square; but if it is not exact, it does not define a function f via $\varphi = \mathrm{d}f$. This means there must be a square around which adding up the values of φ does not vanish. Therefore there must be a square that cannot be built from coordinate squares, i.e. there must be a "hole" in M.

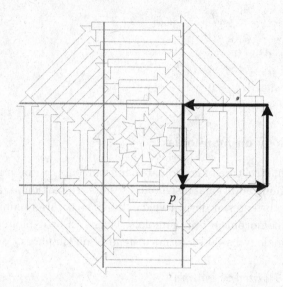

Figure 6.4.1 A closed 1-form must vanish when integrated around any coordinate square, as for example in the square sketched here. For this same 1-form not to be exact, there must be a square around which the integral does not vanish, i.e. a square that is not a coordinate square. Above, the central square is not a coordinate square, since for the depicted 1-form to be smooth, the singular point at the center must be missing from the manifold.

This picture is confirmed and made precise by the Poincaré lemma, which states that if M is contractible, all closed forms are exact. Recall that a contractible space is homotopy equivalent to a point, so that all H_n vanish. Thus the Poincaré lemma says "on a manifold with no holes, closed and exact forms are the same thing." In particular, any point on a manifold is contained in a coordinate chart, which is a contractible neighborhood; thus given a closed form $d\varphi = 0$ and a point p, there is always a neighborhood of p in which we can define ψ such that $\varphi = d\psi$.

6.4.2   *de Rham cohomology*

In the case that there are indeed "holes" present, we can still relate them to closed and exact forms. The **de Rham cohomology** groups are simple to construct; similar to the singular homology groups $H_n(X) \equiv \mathrm{Ker}\partial_n/\mathrm{Im}\partial_{n+1}$, they are the quotient groups $\mathrm{Ker}\, d_n/\mathrm{Im}\, d_{n-1}$, or the closed n-forms modulo the exact n-forms. Thus a de Rham cohomology class is a coset of closed forms that differ by an exact form.

Consider the integral of a closed k-form φ over a k-cycle c in M, i.e. $\int_c \varphi$ where $\mathrm{d}\varphi = 0$ and $\partial c = 0$. Stokes' theorem says that this number is invariant if either c changes by a boundary or φ changes by an exact form:

$$\int_{c+\partial V} \varphi = \int_c \varphi + \int_V \mathrm{d}\varphi = \int_c \varphi$$

$$\int_c \varphi + \mathrm{d}\psi = \int_c \varphi + \int_{\partial c} \psi = \int_c \varphi$$

This integral can thus be viewed as a mapping from a de Rham cohomology coset represented by the closed k-form φ to the real functions on a homology coset represented by the k-cycle c. But this last is just the singular cohomology groups $H^k(M; \mathbb{R})$, so that we have a mapping $\int_c \varphi : H^k_{\text{de Rham}} \to H^k(M; \mathbb{R})$. The **de Rham theorem** states that this mapping is an isomorphism, so that the de Rham and singular cohomology groups with real coefficients are identical for manifolds.

This allows us to deduce information about forms from topological properties. For example, if a manifold M has Betti number $b_k = 0$, then $H^k = 0$ and so every closed k-form on M is exact.

For manifolds, our intuitive picture of n-cycles as "closed surfaces within a space" is quite literal. Every closed oriented submanifold C^k of M^n defines a k-cycle, and a converse is provided by **Thom's theorem**: every k-cycle with real coefficients in M^n is homologous to a real k-chain $\Sigma r_i V_i^k$ of closed oriented submanifolds $V_i^k \subset M^n$.

6.4.3 *Poincaré duality*

For a closed orientable topological manifold M^n, **Poincaré duality** refers to a symmetry that relates the kth homology group to the $(n-k)^{\text{th}}$ group. This symmetry has several consequences:

- There are canonical isomorphisms $H_k \cong H^{n-k}$ for all k
- $H_k \cong H_{n-k}$ modulo their torsion subgroups, i.e. modulo the $\mathbb{Z}$ summands with torsion coefficients $\Leftrightarrow$ in terms of Betti numbers, $b_k = b_{n-k}$
- The torsion subgroups of H_k and H_{n-k-1} are isomorphic
- For a closed non-orientable M, $H_k(M; \mathbb{Z}_2) \cong H^{n-k}(M; \mathbb{Z}_2)$

For orientable non-compact manifolds, a more complicated duality map can also be constructed.

Geometrically, Poincaré duality expresses the existence of dual cell structures on a manifold. For example, in two dimensions the dual is obtained from any cell structure by placing a vertex at the center of every face, with an edge bisecting every original edge. Triangulating the sphere in this way with a regular cell structure yields a polyhedron, and one obtains the dual polyhedron by placing a vertex at the center of each face.

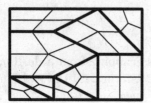 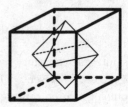

Figure 6.4.2 Dual cell structure in two dimensions for a plane; a cube can be used to triangulate a sphere, as can the dual tetrahedron.

The Platonic solids, i.e. the five convex polyhedra with identical convex regular polygonal faces, are all dual to one another, as the cube is dual to the tetrahedron above. This dual cell structure concept can be generalized to arbitrary dimension, and this can be seen to lead directly to the symmetry between homology on the original cell structure and cohomology on the dual cell structure.

Alexander duality relates homology and cohomology groups for a sphere with a piece deleted: for U a compact locally contractible subspace of S^n, $H_k(S^n - U; \dot{\mathbb{Z}}) \cong H^{n-k-1}(U; \mathbb{Z})$.

Chapter 7

Lie groups

7.1 Combining algebra and geometry

Abstract algebra defines operations on elements, while spaces are defined by relationships between points. Lie groups (pronounced "lee") are an example of a hybrid object, a manifold whose points are also elements in a group. In this chapter we will be primarily concerned with Lie groups, but there are many other useful hybrid algebraic/geometric objects, some of which we briefly mention in this section.

7.1.1 *Spaces with multiplication of points*

Table 7.1.1 Hybrid algebraic/geometric objects.

	Geometric structure	Algebraic structure
H-space	Topological space	Continuous multiplication, identity
Topological group	Hausdorff space	Continuous group operations
Lie group	Differentiable manifold	Differentiable group operations

An **H-space** (AKA Hopf space) is not a group; it may lack inverses or even associativity. An H-space is sometimes defined with $a \mapsto 1a$ and $a \mapsto a1$ only homotopic to the identity, sometimes through basepoint preserving maps. These alternate definitions are equivalent for H-spaces that are cell complexes. The unit vectors in a normed real division algebra have a continuous multiplication and identity, and form the spaces S^1, S^3, and S^7; thus these are H-spaces, and in fact are the only spheres that can be made into H-spaces. S^1 and S^3 are also Lie groups, but S^7 is not even a topological group since it is non-associative.

Since we view manifolds and groups as our most basic geometric and algebraic objects, the structures short of a Lie group have limited interest for us: any group that "looks like a manifold" is automatically a Lie group. More precisely, any group can be made into a topological group, and any topological group that is locally Euclidean can be identified with a single Lie group. The second is because for any topological manifold with continuous group operation, there exists exactly one differentiable structure that turns it into a Lie group.

In any topological group G the **identity component** G^e, the connected component containing the identity, is a normal subgroup. Thus we can view the connected components as equal-sized copies of G^e; in a Lie group, these copies are in fact diffeomorphic. A connected Lie group is sometimes called an analytic group. **Cartan's theorem** states that any closed subgroup of a Lie group is also a Lie group.

7.1.2 *Vector spaces with topology*

In the same way that we defined a topological group to be a space with points that act like group elements, we can define a **topological vector space** to be a Hausdorff space with points that act like vectors over some field, with the vector space operations continuous. However, a better definition might be a vector space with a topology that makes it Hausdorff. This is because the vector space structure already contains topological information in its scalars, which interact as Euclidean spaces. This is reflected in the fact that a finite-dimensional vector space can only be made into a topological vector space with the Euclidean topology; however, this is not true for infinite-dimensional vector spaces.

Taking another approach, we can use norm and inner product structures on a real vector space to turn it into a hybrid object. Recall that a metric space is a topological space with a distance function. A **complete metric space** is one in which the limit of every Cauchy sequence (a sequence of points that become arbitrarily close) is also in the space. In a normed vector space, the norm defines a distance function $\|v - w\|$, which turns the space into a hybrid object, a metric/vector space. A complete normed vector space is called a **Banach space**, and an inner product space that is complete with respect to the norm defined by the inner product is called a **Hilbert space**. All finite-dimensional inner product spaces are automatically Hilbert spaces, but applications in physics typically involve infinite-dimensional spaces, where more care is required.

Introducing the multiplication of vectors, a **Banach algebra** is a Banach space that is an associative algebra satisfying $\|vw\| \leq \|v\| \, \|w\|$. The concept of a conjugate is generalized in a ***-algebra**, a complex associative algebra with an anti-linear mapping * that is both an **involution**, i.e. applied twice to any element it is the identity $v^{**} = v$, and an **anti-automorphism**, i.e. it is an automorphism except under multiplication where we require that $(vw)^* = w^*v^*$. A *-algebra that is also a Banach algebra is called a **B*-algebra**. Finally, if the property $\|vv^*\| = \|v\|^2$ also holds for all vectors, it is called a **C*-algebra**.

The main use of all this in physics is in quantum theories. Hilbert spaces are restrictive enough to act the most like finite-dimensional vector spaces, and the algebra of continuous linear operators on a complex Hilbert space is a C*-algebra. This line of reasoning leads us into analysis, a part of mathematics we will not address in this book; however, here we list some relevant facts for a Hilbert space H, with details omitted:

- Every Hilbert space admits an orthonormal basis (where every element of H is a possibly infinite linear combination of basis vectors)
- Most Hilbert spaces in physics are **separable**, meaning they have a countable dense subset
- All separable Hilbert spaces of countably infinite dimension are isomorphic; thus the references in physics to "Hilbert space"
- Any closed subspace of a Hilbert space has an orthogonal complement
- The dual space H^* (the space of all continuous linear functions from H into the scalars) is isomorphic to H
- Every element of H^* can be written $\langle v, \ \rangle$ for some $v \in H$ (sometimes called the **Riesz representation theorem**)
- This justifies the **bra-ket notation** (AKA Dirac notation), in which we write $|\psi\rangle \in H$, $\langle\psi| \in H^*$, so that in terms of a basis $|\varphi\rangle$ we have $|\psi\rangle = \sum |\varphi_i\rangle \langle\varphi_i|\psi\rangle$

7.2 Lie groups and Lie algebras

Recall that the vector fields on a manifold vect(M) form an infinite-dimensional Lie algebra. The group structure of a Lie group G permits the definition of special vector fields that form a Lie subalgebra of vect(G) with many useful properties. In particular, this special Lie algebra describes the infinitesimal behavior of G, i.e. the behavior near the identity. In physics, Lie groups are used to describe many transformations, with their infinitesimal generators thus described by Lie algebras.

7.2.1 The Lie algebra of a Lie group

Here we define the special vector fields that give Lie groups an associated Lie algebra. The **left translation** mapping $L_g(h) \equiv gh$ is a diffeomorphism on G, as is **right translation** $R_g(h) \equiv hg$. A **left-invariant vector field** A then satisfies

$$\mathrm{d}L_g(A)\,|_h = A|_{L_g(h)}$$

for any g and h. In words, the vector field at any point can be obtained by the left translation of its value at any other point. Thus the vector field is invariant under a left translation diffeomorphism. In particular, a left-invariant vector field is then completely determined by its value at the identity.

The left-invariant vector fields on G under the Lie commutator form its **associated Lie algebra** $\mathfrak{g}$ (which is also isomorphic to the right-invariant vector fields). Since each left-invariant vector field is uniquely determined by its value at the identity element (point) e, $\mathfrak{g}$ is isomorphic to the tangent space $T_e G$, which of course has the same dimension as G. The elements of the Lie algebra $\mathfrak{g} \cong T_e G$ are often called the **infinitesimal generators** of G.

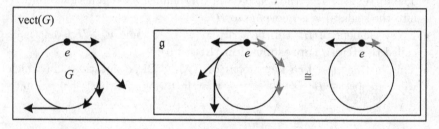

Figure 7.2.1 The Lie group G of rotations of a circle has an associated Lie algebra $\mathfrak{g} \cong \mathbb{R}$.

We can also define **left-invariant forms** by demanding invariance under left translated vector arguments, i.e. we require $L_g^* \varphi = \varphi$ where L_g^* is the pullback. As with left-invariant vectors, left-invariant forms are uniquely determined by their value at the identity. The term **Maurer-Cartan form** can be used to refer to left-invariant 1-forms in general, a particular basis of left-invariant 1-forms, or a $\mathfrak{g}$-valued 1-form that is the identity on left-invariant vector fields.

If we choose a basis of left-invariant 1-forms α^μ (the dual to a basis of $\mathfrak{g}$), we can construct a left-invariant volume form $\alpha^1 \wedge \cdots \wedge \alpha^n$ called a **left Haar measure**. A volume form constructed from a basis of right-invariant 1-forms is called a right Haar measure, and if it is **bi-invariant**, i.e. both left- and right-invariant, it is simply called a Haar measure. Haar measures allow one to construct integrals on G that are invariant under left and/or right translation diffeomorphisms.

7.2.2 The Lie groups of a Lie algebra

Recall that on a differentiable manifold, it is not possible to use a tangent vector v to "transport a point in the direction v" in a coordinate-independent way, since there is no special curve on M among the many that have v as a tangent. On a Lie group this is possible, since the left-invariant vector fields provide a unique flow in the direction of v.

A **one-parameter subgroup** of G is a homomorphism $\phi\colon \mathbb{R} \to G$. Given a left-invariant vector field A, there is a unique one-parameter subgroup ϕ_A such that $\phi_A(0) = e$ and $\dot{\phi}_A(t) = A$ for all t (i.e. $\phi_A(t)$ is the local flow from Section 6.3.2, but being defined for all t it is called simply the **flow** of A). We can then define the **exponential map** exp$\colon \mathfrak{g} \to G$ by
$$\exp(A) \equiv e^A \equiv \phi_A(1).$$
Since scaling the parametrization scales the tangent vectors, we have
$\phi_A(t) = \phi_{tA}(1) = \exp(tA)$.

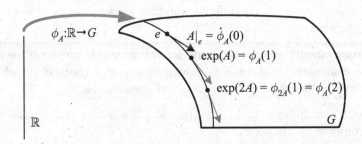

$\phi_A\colon \mathbb{R} \to G$

$A|_e = \dot{\phi}_A(0)$

$\exp(A) = \phi_A(1)$

$\exp(2A) = \phi_{2A}(1) = \phi_A(2)$

$\mathbb{R}$

G

Figure 7.2.2 The exponential map takes $A \in \mathfrak{g}$ to the point a unit distance along its flow.

In particular, the elements of G infinitesimally close to the identity can be written $e + \varepsilon A$. The exponential map is a generalization of familiar exponential functions: if $G = \mathbb{R}^+$, the positive reals under multiplication, $\mathfrak{g} = \mathbb{R}$ and exp is the normal exponential function for real numbers; if G is the non-zero complex numbers under multiplication, $\mathfrak{g} = \mathbb{C}$ and exp is the normal complex exponential function; and if $G = GL(n, \mathbb{R})$, the real invertible $n \times n$ matrices under matrix multiplication, $\mathfrak{g} = gl(n, \mathbb{R})$, the real $n \times n$ matrices, and exp is matrix exponentiation, defined by

$$e^A \equiv \sum_{k=0}^{\infty} \frac{1}{k!} A^k.$$

The multiplication of matrix exponentials does not follow the scalar rule, instead being given by the **Baker-Campbell-Hausdorff formula**:

$$e^A e^B \equiv e^{A+B+\frac{1}{2}[A,B]+\cdots}$$

The terms that continue the series are all expressed in terms of Lie commutators. The terms shown above comprise the entire series if both matrices commute with the commutator, i.e. if $[A, [A, B]] = [B, [A, B]] = 0$. This formula is valid for any associative algebra.

7.2.3   *Relationships between Lie groups and Lie algebras*

The exponential map is a diffeomorphism in some neighborhood of the identity, but in general over G it is neither injective nor surjective. This reflects the fact that in general there are an infinite number of different Lie groups with the same Lie algebra. However, some facts regarding the relationship between a finite-dimensional Lie algebra and its corresponding Lie groups are:

- The exponential map is surjective for any compact connected Lie group
- Any connected Lie group is generated by a neighborhood of the identity, i.e. every element is a finite product of exponentials

The relationship between Lie groups and Lie algebras also extends to derived objects:

- There is a one-to-one correspondence between the connected Lie subgroups of G and the Lie subalgebras of $\mathfrak{g}$
- A connected Lie subgroup of a connected G is normal iff its Lie algebra is an ideal in $\mathfrak{g}$
- The Lie algebra of $G \times H$ is $\mathfrak{g} \oplus \mathfrak{h}$

- Every Lie group homomorphism $\phi\colon G \to H$ determines a Lie algebra homomorphism $d\phi\colon \mathfrak{g} \to \mathfrak{h}$; the converse holds if G is simply connected

Lie algebras can be seen to restrict the topology of Lie groups as compared to general manifolds. For example, a basis of $T_e G$ corresponds to linearly independent left-invariant vector fields on all of G; therefore every Lie group is orientable and parallelizable. The only connected one-dimensional Lie groups are $\mathbb{R}$ and S^1 (under addition of value and angle). Both are abelian, and in fact any connected abelian Lie group is a direct product of these one-dimensional Lie groups. In particular, the only compact 2-dimensional Lie group is the torus $T^2 = S^1 \times S^1$.

7.2.4 *The universal cover of a Lie group*

The relationship between Lie groups and Lie algebras is particularly straightforward for simply connected Lie groups:

- Every Lie algebra corresponds to a unique simply connected Lie group G^*
- There is a group homomorphism ϕ from this unique simply connected Lie group G^* to any other connected Lie group G with the same Lie algebra, with $\mathrm{Ker}\phi \cong \pi_1(G)$ discrete

This last implies that for any Lie group $G \cong G^*/\pi_1(G)$, the simply connected Lie group G^* with the same Lie algebra has a fixed number of points that map down to any point in G. G^* can thus be pictured as "wrapping around" any such G some number of times, and is therefore called the **universal covering group**.

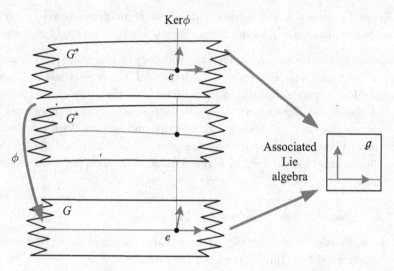

Figure 7.2.3 A depiction of the universal covering group G^* and its homomorphism to any other Lie group G with the same Lie algebra. A one-dimensional subalgebra and corresponding one-dimensional subgroups are shown as lines.

The idea of a space covering another generalizes to any topological space: a **covering space** C of a space X has a continuous surjective map to X whose inverse in a neighborhood of any point of X is a union of mutually disjoint open sets homeomorphic to that neighborhood. The points that map to a point $p \in X$ are called the **fiber** over p, and the disjoint open sets over a neighborhood of p are called **sheets**. Under reasonable connectivity requirements, every space then has a unique simply connected **universal covering space** that covers all connected covers.

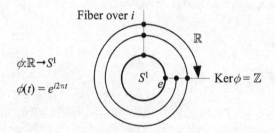

Figure 7.2.4 The infinite-sheeted universal covering space $\mathbb{R}$ of S^1.

7.3 Matrix groups

The most common Lie groups are the **matrix groups** (AKA linear groups), which are Lie subgroups of the group of real or complex $n \times n$ invertible matrices, denoted $GL(n, \mathbb{R})$ and $GL(n, \mathbb{C})$. We can also consider the **linear groups**, which are Lie subgroups of $GL(V)$, the group of invertible linear transformations on a real or complex vector space V. One can then choose a basis of V to get a (non-canonical) isomorphism to the matrix group, e.g. from $GL(\mathbb{R}^n)$ to $GL(n, \mathbb{R})$.

$\triangle$ The distinction between the abstract linear groups and the basis-dependent matrix groups is not always made, and the notation is used interchangeably. Alternative notation includes $GL_n(\mathbb{R})$, and the field and/or dimension is often omitted, yielding notation such as GL_n, $GL(n)$, $GL(\mathbb{R})$, or GL.

These groups can be seen to be Lie groups by taking global coordinates to be the real matrix entries or the real components of the complex entries. Thus $GL(n, \mathbb{R})$ is a manifold of dimension n^2, and $GL(n, \mathbb{C})$ has manifold dimension $2n^2$. Any subgroup of GL that is also a submanifold is then automatically a Lie subgroup.

We can also consider Lie groups defined by invertible matrices with entries in $\mathbb{H}$ or $\mathbb{O}$, since even though they cannot be viewed as linear transformations on a vector space, they still form a group and are manifolds with respect to the real components of their entries.

$\triangle$ Some matrix groups can also be viewed as a **complex Lie group**, a group that is also a complex manifold. For example, $GL(n, \mathbb{C})$ can be viewed as an n^2-dimensional complex Lie group instead of as a real Lie group of dimension $2n^2$. It is important to distinguish between a complex Lie group and a real Lie group defined by matrices with complex entries.

7.3.1 Lie algebras of matrix groups

The Lie algebra associated with a matrix group is denoted by the same abbreviation as the Lie group, but with lowercase letters; e.g. the Lie algebra of $GL(n, \mathbb{R})$ is denoted $gl(n, \mathbb{R})$. $gl(n, \mathbb{R})$ is easily seen to be the set of all real $n \times n$ matrices under the Lie commutator, and in general the Lie algebra associated with a matrix group can be expressed as matrices with entries in the same division algebra as the matrix group.

> △ It is important to remember that the multiplication operation on the matrices of a Lie algebra is that of the Lie commutator using matrix multiplication.

If an element of $GL(n, \mathbb{R})$ is considered to be a linear transformation on $\mathbb{R}^n$, an element of $gl(n, \mathbb{R})$ is an infinitesimal generator of a linear transformation. Thus an element A of $gl(n, \mathbb{R})$ can be viewed as a vector field on $\mathbb{R}^n$ that "points in the direction of a linear transformation," i.e. as a matrix it linearly transforms a vector v into the tangent to the path in $\mathbb{R}^n$ traced by the one-parameter subgroup $\phi_A(t)$ applied to v.

> △ This view of the element A as a vector field on $\mathbb{R}^n$ should not be confused with the view of gl as a vector field on GL; incorporating this latter view would make gl a "vector field of vector fields."

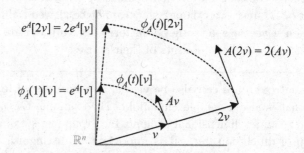

Figure 7.3.1 An element of $GL(n, \mathbb{R})$ is a linear transformation on $\mathbb{R}^n$, while an element A of the associated Lie algebra $gl(n, \mathbb{R})$ is a vector field on $\mathbb{R}^n$ that "points in the direction" of the element $e^A \in GL(n, \mathbb{R})$.

7.3.2   *Linear algebra*

Here we recall some basics of linear algebra, which is assumed to be familiar to the reader. We start by collecting some terminology:

- A^{T}: **transpose** of A, reflecting entries across the diagonal
- $A^{\mathrm{T}} = -A$: **anti-symmetric** (AKA skew-symmetric) matrix
- $A^{\dagger}$: **adjoint** (AKA hermitian conjugate) of a matrix, the transposed complex conjugate (also denoted A^*)
- $A^{\dagger} = A$: **hermitian** matrix, a matrix that is **self-adjoint**
- $A^{\dagger} = -A$: **anti-hermitian** (AKA skew-hermitian) matrix $\Rightarrow iA$ is hermitian
- $A^{\dagger}A = I$: **unitary (orthogonal)** matrix for complex (real) entries
- $A^{\dagger}A = AA^{\dagger}$: **normal** matrix, e.g. a hermitian or unitary matrix
- **Eigenvalues**: scalars a such that $Av = av$ for vectors v, the **eigenvectors**
- $\mathrm{tr}(A)$: the **trace** of the matrix A, the sum of the diagonal entries
- $\det(A)$: determinant of A
- **Singular** means $\det(A) = 0$, **unimodular** can mean either $|\det(A)| = 1$ or $\det(A) = 1$
- **Similarity transformation**: $A \to BAB^{-1}$ by a nonsingular matrix B

Some basic facts are:

- A similarity transformation $A \to BAB^{-1}$ is equivalent to a change of the basis defining the vector components operated on by A, where the change of basis has matrix B^{-1} so that $v \to Bv$
- The eigenvalues, determinant and trace of A are independent of basis $\Rightarrow$ unchanged by a similarity transformation
- The trace is a cyclic linear map: $\mathrm{tr}(ABC) = \mathrm{tr}(BCA) = \mathrm{tr}(CAB)$
- The determinant is a multiplicative map: $\det(rAB) = r^n \det(A)\det(B)$
- The trace equals the sum of eigenvalues; the determinant equals their product
- $\det(\exp(A)) = \exp(\mathrm{tr}(A))$; $(\exp(A))^{\dagger} = -\exp(A^{\mathrm{T}})$
- $\det(I + \varepsilon A) = 1 + \varepsilon \mathrm{tr} A + \ldots$
- A hermitian matrix has real eigenvalues and orthogonal eigenvectors
- **Spectral theorem**: any normal matrix can be diagonalized by a unitary similarity transformation

As previously noted, we can geometrically interpret an element of a matrix group with real entries as a transformation on $\mathbb{R}^n$. Such a transformation preserves the orientation of $\mathbb{R}^n$ if its determinant is positive, and preserves volumes if the determinant has absolute value one.

Any bilinear form φ on $\mathbb{R}^n$ can be represented by a matrix in the standard basis, with the form operation then being $\varphi(v, w) = v^{\mathrm{T}} \varphi w$. The

group of matrices that preserve a form φ consists of matrices A that satisfy $\varphi\left(Av, Aw\right) = \varphi\left(v, w\right) \Leftrightarrow (Av)^{\mathrm{T}} \varphi(Aw) = v^{\mathrm{T}} \varphi w \Leftrightarrow A^{\mathrm{T}} \varphi A = \varphi$. Any similarity transformation simply changes the basis of each A, leaving the group of matrices that preserve the form unchanged. Thus we can concern ourselves only with a canonical form of the preserved form. In $\mathbb{R}^n$, we have several naturally defined forms:

- The Euclidean inner product, with canonical form I
- The pseudo-Euclidean inner product of signature (r, s), with canonical form $(r + s = n)$

$$\eta = \begin{pmatrix} I_r & 0 \\ 0 & -I_s \end{pmatrix}$$

- The symplectic form, with canonical form

$$J = \begin{pmatrix} 0 & I_{n/2} \\ -I_{n/2} & 0 \end{pmatrix}$$

Any matrix group defined as preserving one of these canonical forms then preserves all forms in the corresponding similarity class. Some matrix groups with entries in $\mathbb{C}$ or $\mathbb{H}$ can also be viewed as preserving a form in the vector space $\mathbb{C}^n$ or module $\mathbb{H}^n$, but we will mainly view these as linear transformations on $\mathbb{R}^{2n}$ or $\mathbb{R}^{4n}$.

7.3.3 Matrix groups with real entries

Here we summarize some of the common matrix groups with real entries, with a focus on their geometrical properties as linear transformations on $\mathbb{R}^n$.

Table 7.3.1 Matrix groups with real entries.

Name	Geometry	Matrix	Lie algebra
$GL(n,\mathbb{R})$ General linear group	Arbitrary change of basis in $\mathbb{R}^n$	$n \times n$ matrices with $\det(A) \neq 0$	All $n \times n$ matrices
$GL(n,\mathbb{R})^e$	Preserves orientation	$\det(A) > 0$	$gl(n,\mathbb{R})$
$SL(n,\mathbb{R})$ Special linear group	Preserves orientation and volume	$\det(A) = 1$	$\operatorname{tr}(A) = 0$
$O(n)$ Orthogonal group	Preserves the Euclidean inner product: rotations and reflections	$A^T A = I$ $\Rightarrow \det(A) = \pm 1$	$A^T = -A$
$SO(n)$ Special orthogonal group	Proper rotations (preserves orientation)	$A^T A = I$, $\det(A) = 1$	$o(n,\mathbb{R})$
$O(r,s)$ Pseudo-orthogonal group	Preserves the pseudo-Euclidean inner product	$A^T \eta A = \eta$ $\Rightarrow \det(A) = \pm 1$	Matrices ηA for A anti-symmetric
$SO(r,s)$ Special pseudo-orthogonal group	As above, but preserves orientation	$A^T \eta A = \eta$, $\det(A) = 1$	$o(r,s)$
$Sp(2n,\mathbb{R})$ Real symplectic group	Preserves the symplectic form	$A^T J A = J$ $\Rightarrow \det(A) = 1$	$JA + A^T J = 0$

Notes: Just as $GL(n,\mathbb{R})$ is often written GL_n, similar notation is sometimes used for other groups. The notation does not distinguish between abstract and matrix groups; we will attempt to note the distinction when relevant. $GL(n,\mathbb{R})^e$ is often written GL_n^+ or similar. An immediate result from their definitions is $O(r,s) \cong O(s,r)$ and $SO(r,s) \cong SO(s,r)$. The notation $Sp(2n,\mathbb{R})$ reflects the fact that J only exists for even-dimensional matrices; however, sometimes it is denoted $Sp(n,\mathbb{R})$, where the group still consists of $2n \times 2n$ matrices. We will always use notation consistent with the size of the defining matrices.

7.3.4 Other matrix groups

Here we summarize some other common matrix groups. We again stress that although they are defined in terms of matrices with non-real elements, these are all real Lie groups.

Table 7.3.2 Other matrix groups.

Name	Matrix	Lie algebra
$GL(n, \mathbb{C})$	$n \times n$ complex matrices with $\det(A) \neq 0$	All complex $n \times n$ matrices
$SL(n, \mathbb{C})$	$\det(A) = 1$	$\operatorname{tr}(A) = 0$
$U(n)$ Unitary group	$A^\dagger A = AA^\dagger = I$ $\Rightarrow |\det(A)| = 1$	$A^\dagger = -A$
$SU(n)$ Special unitary group	$A^\dagger A = AA^\dagger = I,$ $\det(A) = 1$	$A^\dagger = -A,$ $\operatorname{tr}(A) = 0$
$Sp(2n, \mathbb{C})$ Complex symplectic group	$A^{\mathrm{T}} J A = J$	$JA + A^{\mathrm{T}} J = 0$
$Sp(n)$ Quaternionic symplectic group	$n \times n$ quaternionic matrices with $A^\dagger A = AA^\dagger = I$ where $\dagger$ uses the quaternionic conjugate	$A^\dagger = -A$

Notes: $U(n)$ is the complex version of $O(n)$, and can be viewed as preserving the standard inner product $\langle v, w \rangle \equiv v^\dagger w$ on $\mathbb{C}^n$; however it does not form a complex Lie group. The complex versions of the pseudo-orthogonal and special pseudo-orthogonal groups can be similarly defined.

The quaternionic symplectic group $Sp(n)$ is also called the quaternionic unitary group, which better matches the definition above. An equivalent definition is $Sp(n) \equiv U(2n) \cap Sp(2n, \mathbb{C})$, and thus $Sp(n)$ is also called the unitary symplectic group. Unlike in the real and complex cases, it is also compact, and so yet another term used is the compact symplectic group. One can also view the relationships between the three symplectic groups in terms of their Lie algebras; this will be seen in Table 7.5.1 in Section 7.5.2.

Additional matrix groups can be defined by generalizing more of the above constructions to mixed signatures and quaternionic entries, but they are not as frequently used in physics and we will not cover them here.

7.3.5 *Manifold properties of matrix groups*

As real manifolds, we can list various properties of the matrix groups.

Table 7.3.3 Manifold properties of matrix groups for $n > 1$ and $rs \neq 0$.

Group	Dimension	Compact	Connectedness
$GL(n, \mathbb{R})$	n^2	No	2 components
$SL(n, \mathbb{R})$	$n^2 - 1$	No	Connected
$O(n)$	$n(n-1)/2$	Yes	2 components
$SO(n)$	$n(n-1)/2$	Yes	Connected
$O(r, s)$	$n(n-1)/2$	No	4 components
$SO(r, s)$	$n(n-1)/2$	No	2 components
$Sp(2n, \mathbb{R})$	$(2n)(2n+1)/2$	No	Connected
$GL(n, \mathbb{C})$	$2n^2$	No	Connected
$SL(n, \mathbb{C})$	$2(n^2 - 1)$	No	Simply connected
$U(n)$	n^2	Yes	Connected
$SU(n)$	$n^2 - 1$	Yes	Simply connected
$Sp(2n, \mathbb{C})$	$(2n)(2n+1)$	No	Simply connected
$Sp(n)$	$n(2n+1)$	Yes	Simply connected

Notes: In particular, since $U(n)$ is compact and connected, any unitary matrix U can be written as $U = e^{iH}$ for some hermitian matrix H.

Since the exponential map is surjective for any compact connected Lie group, it is surjective for $SO(n)$, $U(n)$ and $SU(n)$. As it turns out, it is also surjective for $SO(3,1)^e$ and $GL(n, \mathbb{C})$.

We can also characterize the topology of matrix groups by noting some diffeomorphisms of their manifolds:

- $O(n+1)/O(n) \cong SO(n+1)/SO(n) \cong S^n$
- $U(n+1)/U(n) \cong SU(n+1)/SU(n) \cong S^{2n+1}$
- In particular, we then have $U(1) \cong SO(2) \cong S^1$, $SU(2) \cong S^3$
- $O(n) \cong S^0 \times SO(n)$; $U(n) \cong S^1 \times SU(n)$
- Thus $SO(n+1)/O(n) \cong \mathbb{R}\mathrm{P}^n$; $SU(n+1)/U(n) \cong \mathbb{C}\mathrm{P}^n$
- In particular, we then have $SO(3) \cong \mathbb{R}\mathrm{P}^3$
- $SO(4) \cong S^3 \times SO(3)$; $SO(8) \cong S^7 \times SO(7)$
- $Sp(2, \mathbb{R}) \cong SL(2, \mathbb{R})$; $Sp(2, \mathbb{C}) \cong SL(2, \mathbb{C})$; $Sp(1) \cong SU(2) \cong S^3$

With regard to homotopy groups, some facts are:

- $\pi_1(G)$ is abelian for any Lie group G (in fact for any H-space)
- $\pi_2(G) = 0$ for any Lie group G
- $\pi_1(SO(n)) = \mathbb{Z}_2$ for $n > 2$; $\pi_3(SU(n)) = \mathbb{Z}$ for $n > 1$

7.3.6 Matrix group terminology in physics

An important fact used in physics is that $SU(2)$ is the universal covering group of $SO(3)$. For example, the four complex numbers associated with an element of $SU(2)$ are called **Cayley-Klein parameters**, and via this homomorphism can be used to specify a proper rotation in $\mathbb{R}^3$, i.e. an element of $SO(3)$. $SU(2)$ is a double cover, so there are two Cayley-Klein parameters corresponding to every proper rotation.

The connected components of matrix groups are usually related to determinant signs; we can see this by studying $O(3,1)$, which in physics is called the (homogeneous) **Lorentz group**. It is of dimension 6, and consists of rotations and reflections in **Minkowski space** (AKA spacetime), $\mathbb{R}^4$ with the Minkowski metric, where the positive signatures correspond to space and the negative signature to time. $O(3,1)$ has 4 connected components, corresponding to whether the orientation of time and/or space is reversed. The **proper Lorentz group** $SO(3,1)$ consists of the identity component and the connected component of transformations that reverse both space and time, while the **orthochronous Lorentz group** consists of the two components that preserve the orientation of time. The identity component $SO(3,1)^e$ is then called the **proper orthochronous Lorentz group** (AKA restricted Lorentz group). A "rotation" by an angle ϕ in a "time-like plane" which includes vectors with negative lengths is called a **Lorentz boost** of **rapidity** ϕ.

The **Poincaré group** (AKA inhomogeneous Lorentz group) $IO(3,1)$ is the semidirect product of the translations in 4 possible directions with $O(3,1)$; the product is semidirect since the translations are a normal subgroup and every element can be written in exactly one way as a translation followed by a rotation. It has dimension 10, or $n(n+1)/2$ in general. The above adjectives can also be applied to the Poincaré group. All of the above analysis was for the "mostly pluses" signature $(3,1)$, but the results also hold for the "mostly minuses" signature $(1,3)$.

Similarly, the **Euclidean group** $E(n) \equiv \mathbb{R}^n \rtimes O(n)$ is the semidirect product of translations in $\mathbb{R}^n$ with $O(n)$, while the **special Euclidean group** $SE(n)$ takes the semidirect product with $SO(n)$. These are subgroups of the **affine group** $Aff(n,\mathbb{R}) \equiv \mathbb{R}^n \rtimes GL(n,\mathbb{R})$ and its identity component the **special affine group**, respectively. All of these "inhomogeneous" groups, i.e. groups formed by semidirect products whose elements are a translation v followed by a linear transformation A, can be viewed as matrix groups of the form

$$\begin{pmatrix} A & v \\ 0 & 1 \end{pmatrix},$$

where v is a column vector and the matrices can be considered as acting on vectors in **homogeneous coordinates** (AKA projective coordinates), in which a component 1 is appended:

$$\begin{pmatrix} A & v \\ 0 & 1 \end{pmatrix} \begin{pmatrix} w \\ 1 \end{pmatrix} = \begin{pmatrix} Aw + v \\ 1 \end{pmatrix}$$

$\triangle$ Note that in homogeneous coordinates scalar multiples of vectors are identified, e.g. $(1, 2, 3, 1) = (2, 4, 6, 2)$. A possible source of confusion is that inhomogeneous groups act on vectors in homogeneous coordinates.

7.4 Representations

The group of three-dimensional rotations $SO(3)$ can "act" on various objects, for example the space $\mathbb{R}^3$ or the unit sphere S^2. These example actions are "symmetries" of the objects, in that the rotated objects are isomorphic to the original, with each element of $SO(3)$ thus "represented" by an automorphism of the object. These ideas are formalized by actions and representations, which are homomorphisms from each element of an algebraic object to a morphism from a space to itself. Group representations in particular are heavily used in physics.

Note that a matrix group (or algebra) has a **defining representation** (AKA standard representation) on the space of its entries, e.g. $\mathbb{R}^n$ or $\mathbb{C}^n$, but in a given situation one may be working with a different representation. The defining representation of a Lie group is also often called the **fundamental representation**, but this term has a different meaning when used in the classification of Lie algebras.

$\triangle$ It is important to keep in mind which vector space is meant in a given situation; e.g. in the context of a representation of $SO(n)$ acting on an object in $\mathbb{R}^m$, there is the space $\mathbb{R}^n$ used to define the group, the possibly different Euclidean space $\mathbb{R}^m$ the representation is acting on, and the space $\mathbb{R}^{n(n-1)/2}$ that the charts of the manifold of the Lie group map to.

7.4.1   *Group actions*

A **group action** (AKA left action, realization, representation) of any group G on any set X is a homomorphism $\rho\colon G \to \mathrm{Aut}\,(X)$, where $\mathrm{Aut}\,(X)$ is the group of automorphisms of X (the symmetry group). Since in most applications a group action is fixed, we will write the action of $g \in G$ on $x \in X$ as simply $g(x)$, or $g_\rho\,(x)$ if the homomorphism needs to be made explicit; other common notations are gx, $\rho_g(x)$ and $\rho(g)(x)$. When acted on by G, X is sometimes called a **G-set** (or a **G-space** if it is a space).

$\triangle$ As we will see, some specific types of actions (e.g. reps) may be required to preserve additional structures that exist on X (e.g. a vector space structure), but in general, the automorphism group is that of X as a set or space, with any additional structure (e.g. that of a fiber bundle) disregarded.

Since a left action is a homomorphism, $g \circ h$ and gh are required to be the same automorphism, i.e.

$$g\,(h\,(x)) = (gh)\,(x) \;\; \forall x \in X.$$

A **right action** operates from the right within the group, and so instead requires that

$$g\,(h\,(x)) = (hg)\,(x)\,,$$

and is often written xg. Note that a left action can be turned into a right action (and vice versa) via the inverse; e.g. if G has a left action and we define a new action $g_R(x) \equiv g^{-1}(x)$, then

$$
\begin{aligned}
g_R\,(h_R\,(x)) &= g^{-1}\left(h^{-1}(x)\right) \\
&= (g^{-1}h^{-1})(x) \\
&= (hg)^{-1}(x) \\
&= (hg)_R\,(x)\,.
\end{aligned}
$$

Some definitions related to a group action are:

- **Orbit** of $x \in X$: $\mathrm{orbit}(x) \equiv \{g(x) \mid g \in G\}$; i.e. all points of X that can be reached from x by the action of some g
- **Isotropy group** (AKA little group, stabilizer subgroup) of x: the subgroup $I\,(x) \equiv \{g \mid g\,(x) = x\}$; i.e. all elements of G that leave x fixed
- **Transitive** action: $\forall x, y \; \exists g \mid y = g\,(x) \Leftrightarrow X$ is a single orbit; i.e. any two points are related by the action of some g

- **Faithful** (AKA effective) action: $\forall g \neq h \; \exists x \mid g(x) \neq h(x) \Leftrightarrow \rho$ is injective; i.e. every g is mapped to a distinct automorphism
- **Free** (AKA semiregular, fixed point free) action: $\forall g \neq h, \; g(x) \neq h(x) \forall x \Leftrightarrow$ only e has a fixed point; i.e. the orbit of every x is an injective map of G
- **Regular** (AKA simply transitive, sharply transitive) action: $\forall x, y \; \exists$ unique $g \mid y = g(x) \Leftrightarrow$ transitive and free; i.e. any two points are related by the action of one g

One can state various relationships between these properties, for example: free implies faithful; free is equivalent to all isotropy groups being trivial; and G acts transitively on any orbit of X. If the action of G is transitive, then X is called a **homogeneous space** for G; if the action is also free (i.e. regular), then X is called a **principal homogeneous space** or **G-torsor**. A G-torsor is isomorphic to G as a set or space, but there is no uniquely defined identity element; it can thus be thought of as a group "with the identity forgotten." The action of G on itself by left or right multiplication is regular.

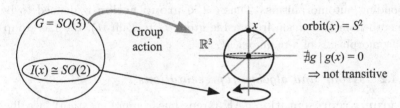

Figure 7.4.1 The action of the three dimensional rotations $SO(3)$ on $\mathbb{R}^3$ is not transitive, since two points at different radii cannot be reached from each other by the action of a rotation; is faithful, since every rotation is a distinct automorphism; but is not free, since every rotation leaves an axial line fixed. The orbit of x is the sphere of the same radius, and the isotropy group of x is the two dimensional rotations around the axis it determines.

If a group G has a left action on two sets X and Y, a mapping $f : X \to Y$ is called **equivariant** if

$$f(g(x)) = g(f(x))$$

for all g and x. In other words, an equivariant map is a homomorphism with respect to the group action; it is therefore also sometimes called a **G-map** or **G-homomorphism**. This definition has to be modified if we extend it to right actions, where we take advantage of the property $(gh)^{-1} = h^{-1}g^{-1}$ to maintain ordering:

Table 7.4.1 The equivariance condition for a map $f \colon X \to Y$ between two G-sets, using two common notations.

	Left action on Y	Right action on Y
Left action on X	$f\left(g\left(x\right)\right) = g\left(f\left(x\right)\right)$ $f(gx) = gf(x)$	$f\left(g\left(x\right)\right) = g^{-1}\left(f\left(x\right)\right)$ $f(gx) = f(x)g^{-1}$
Right action on X	$f\left(g\left(x\right)\right) = g^{-1}\left(f\left(x\right)\right)$ $f(xg) = g^{-1}f(x)$	$f\left(g\left(x\right)\right) = g\left(f\left(x\right)\right)$ $f(xg) = f(x)g$

If G has a left action on X, and we denote the left cosets of the isotropy group as $G/I(x)$, then the map $f \colon G/I(x) \to \mathrm{orbit}(x)$ defined by $gI(x) \mapsto gx$ is equivariant. The **orbit-stabilizer theorem** states that this map is also bijective. Such a map is sometimes called a G-map isomorphism. For finite G, the corollary $|G| \, / \, |I(x)| = |G : I(x)| = |\mathrm{orbit}(x)|$ is also sometimes referred to as the orbit-stabilizer theorem, where $|\mathrm{orbit}(x)|$ denotes the number of elements in the set.

A Lie group has the additional structure of a differentiable manifold, which is required to carry over the action homomorphism to the corresponding automorphisms. Thus a **Lie group action** is defined to be a smooth homomorphism from a Lie group G to $\mathrm{Diff}(M)$, the Lie group of diffeomorphisms of a manifold M.

7.4.2 *Group and algebra representations*

A **group representation** (AKA rep, linear representation) is a linear group action on a real or complex vector space V, i.e. a homomorphism $\rho \colon G \to GL(V)$ from G to the Lie group of linear invertible automorphisms of the manifold $V = \mathbb{R}^n$ or $\mathbb{C}^n$. We can choose a basis of V to get a isomorphism from $GL(V)$ to $GL(n, \mathbb{R})$ or $GL(n, \mathbb{C})$, in which case the representation is called a **matrix representation**. For a matrix rep, the transpose switches left and right actions like the inverse, as does an action by the matrix on row vectors instead of column vectors; in particular, a left matrix rep on vector components $v'^{\mu} = g^{\mu}{}_{\lambda} v^{\lambda}$ is equivalent to a left action

$$e'_{\mu} = (g^{-1})^{\lambda}{}_{\mu} e_{\lambda} = [e_1 \cdots e_n] \left[g^{-1}\right]$$

on the basis, since the inverse of the matrix acts on a row vector. The G-space V is called a **representation space**, and an equivariant map between representation spaces of the same group is called an **intertwiner** (AKA intertwining map); note that by the definition of equivariant, an intertwiner is itself a linear map.

> △ It is common to use "representation" to refer to the representation
> space V, with the group G and the mapping ρ inferred from context.

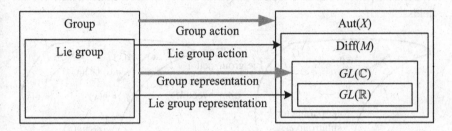

Figure 7.4.2 Actions and representations of groups and Lie groups. The mappings are homomorphisms for groups, smooth homomorphisms for Lie groups.

Now, a group representation $\rho\colon G \to GL(V)$, being a homomorphism, satisfies $\rho\,(gh) = \rho\,(g)\,\rho\,(h)$. Similarly, an **algebra representation** of an associative algebra $\mathfrak{a}$ is defined to be a linear homomorphism $\rho\colon \mathfrak{a} \to gl\,(V)$, e.g. for scalar a and vectors A, B, C in $\mathfrak{a}$, we require that $\rho\,(aA + BC) = a\rho\,(A) + \rho\,(B)\,\rho\,(C)$. An algebra representation is also referred to as a **G-module**, or just a module, since if we ignore scalars in both V and $\mathfrak{a}$, V acted on by $\mathfrak{a}$ can be viewed as a module with vectors in the abelian group V and scalars in the ring $\mathfrak{a}$.

We can now use the Lie commutator to define the related **Lie algebra representation** of a Lie algebra $\mathfrak{g}$ as a smooth linear homomorphism $\rho\colon \mathfrak{g} \to gl\,(V)$, i.e. we require $\rho\,([A, B]) = \rho\,(A)\,\rho\,(B) - \rho\,(B)\,\rho\,(A)$. Note that a Lie algebra derived from a real Lie group is by definition a real vector space, since it lies in the tangent space of a real manifold; thus the scalar field in such a Lie algebra, even if defined by complex matrices, is the field of reals.

7.4.3   Lie group and Lie algebra representations

A Lie group action ρ is a smooth homomorphism from G to Diff(M). An element of G near the identity then moves each point of M to a nearby point. So for any vector $A \in \mathfrak{g}$, the one-parameter subgroup ϕ_A from the identity along A maps to a curve in M at each point. The differential of this mapping takes A to a vector field on M, and this relation is in fact a Lie algebra homomorphism $d\rho\colon \mathfrak{g} \to \text{vect}(M)$, the corresponding Lie algebra action. Recalling the construction of the Lie derivative, we see that the Lie algebra action of $A \in \mathfrak{g}$, called the **fundamental vector field** corresponding to A, is the vector field on M whose local flow is the Lie group action of the one-parameter subgroup ϕ_A. If G acts on itself by right translation, the fundamental vector fields are just the left-invariant vector fields.

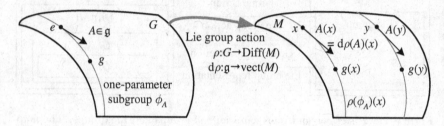

Figure 7.4.3 The corresponding Lie algebra action obtained from a Lie group action.

In the case of a Lie group representation on a real or complex vector space V, the corresponding Lie algebra representation maps $\mathfrak{g}$ to a linear subalgebra of vect(V) that is isomorphic to $gl(V)$. The Lie bracket in this case is the Lie commutator, whether viewed as that of vector fields, of transformations, or of matrices. Similarly one can show that if a Lie algebra $\mathfrak{g}$ has a matrix representation, and a compact connected Lie group G corresponds to $\mathfrak{g}$, then G has a matrix representation given by the matrix exponential of the Lie algebra representation.

Every finite-dimensional real Lie algebra has a faithful finite-dimensional real representation, i.e. can be viewed as a class of real matrices. This result is a special case of two theorems dealing with scalars in more general fields, **Ado's theorem** and **Iwasawa's theorem**. The analog is not true in general for finite-dimensional Lie groups, although most Lie groups used in physics can be viewed as matrix groups. A standard counter-example given is the universal covering group of $SL(2, \mathbb{R})$, which is infinite-sheeted and therefore has no faithful finite-dimensional representation.

7.4.4 *Combining and decomposing representations*

If G and H are groups or Lie groups with representations on vector spaces V and W, we can define the direct sum of the representations as the representation of $G \times H$ on $V \oplus W$ defined by $(g, h)(v, w) \equiv (g(v), h(w))$. The

Lie algebra of $G \times H$ is $\mathfrak{g} \oplus \mathfrak{h}$, and it then has a representation on $V \oplus W$ similarly given by $(A, B)(v, w) \equiv (A(v), B(w))$.

Since every linear transformation leaves the origin invariant, no linear representation is transitive. However, one can ask that at least no vector subspace be invariant. An **irreducible linear representation** (AKA irrep,) on V is defined as a group or algebra representation that has no non-trivial **invariant subspace** (AKA subrepresentation, or submodule if an algebra rep) $0 \subset W \subset V$ such that $gW \subset W \, \forall g \in G$. A representation is **completely reducible** (AKA decomposable) if the orthogonal complement of every invariant subspace is also invariant; any finite-dimensional completely reducible representation can then be written as a direct sum of irreps. Referring back to Figure 7.4.1, the action of $SO(2)$ on $\mathbb{R}^3$ is completely reducible, and can be written as the direct sum of the identity irrep on the axis of rotation $\mathbb{R}^1$ and the rotation irrep on the plane $\mathbb{R}^2$ orthogonal to it.

Note that a representation can be reducible but not completely reducible, i.e. can have an invariant subspace and yet not be a direct sum of irreps. However, most representations of interest are either irreducible or completely reducible:

- Every representation of a finite group is completely reducible
- Every representation of a compact Lie group is completely reducible
- Every unitary representation is completely reducible
- **Weyl's theorem:** every representation of a Lie algebra is completely reducible iff the Lie algebra is semisimple (semisimple will be defined in Section 7.5.1)
- Every representation of a connected semisimple Lie group is completely reducible

Again considering groups or Lie groups G and H with representations on V and W, we can define the tensor product of the representations as the representation of $G \times H$ on $V \otimes W$ defined by $(g, h)(v \otimes w) \equiv g(v) \otimes h(w)$. In this case the representation of the Lie algebra $\mathfrak{g} \oplus \mathfrak{h}$ is given by $(A, B)(v \otimes w) = A(v) \otimes I + I \otimes B(w)$, in order to make it linear on $V \otimes W$.

The tensor product of two representations of the same group G can be viewed as a new representation of G on the vector space $V \otimes W$ given by $g(v \otimes w) \equiv g(v) \otimes g(w)$. Even if the two original representations are irreducible, this new tensor product representation may not be; decomposing it into a direct sum of irreps is called **Clebsch-Gordan** theory.

By noting that the kernel and image of an intertwiner are invariant subspaces, one arrives at **Schur's Lemma**, which states that any intertwiner between irreps is either zero or an isomorphism. This has several immediate consequences, which are sometimes referred to themselves as Schur's Lemma:

- Any self-intertwiner of a finite-dimensional complex irrep is a multiple of the identity map
- Any two intertwiners between finite-dimensional complex irreps differ by only a complex constant multiple
- Any matrix in the center of the image of a complex irrep is a multiple of the identity matrix
- A complex irrep maps any element in the center of a Lie group to a multiple of the identity transformation
- Any irrep of an abelian Lie group is one-dimensional (as a manifold)

For a real Lie algebra $\mathfrak{g}$, we can consider its complexification $\mathfrak{g}_{\mathbb{C}}$ as a complex Lie algebra. We can then ask, if $\mathfrak{g}_{\mathbb{C}}$ has an irrep on $\mathbb{C}^n$ (i.e. as an algebra with complex matrices in $\mathbb{C}(n)$ as vectors and scalars in $\mathbb{C}$), does this correspond to an irrep of $\mathfrak{g}$ on $\mathbb{C}^n$ (i.e. as an algebra of complex matrices in $\mathbb{C}(n)$ as vectors and scalars in $\mathbb{R}$)? The answer is yes; the irreps on $\mathbb{C}^n$ of $\mathfrak{g}$ are one to one with those of $\mathfrak{g}_{\mathbb{C}}$ as a complex Lie algebra.

7.4.5 *Other representations*

For a Lie group G, the inner automorphism $\phi_g : G \to G$ induced by a fixed $g \in G$ is defined by $h \mapsto ghg^{-1}$, and can be viewed as an action of G on itself. At the identity e we then have the map $(\mathrm{d}\phi_g)'|_e : \mathfrak{g} \to \mathfrak{g}$. The **adjoint representation** $\mathrm{Ad} : G \to GL(\mathfrak{g})$ represents G on $\mathfrak{g}$, and for $A \in \mathfrak{g}$ is defined by $g_{\mathrm{Ad}}(A) \equiv (\mathrm{d}\phi_g)|_e(A)$; it is often denoted $\mathrm{Ad}_g A$. Using the exponential map, one can show that $\exp(t g_{\mathrm{Ad}}(A)) = g \exp(tA) g^{-1}$. If G is a matrix group, so that g and A are both matrices, the adjoint representation is simply the similarity transformation $g_{\mathrm{Ad}}(A) = gAg^{-1}$.

The adjoint representation also sometimes refers to the representation of the Lie algebra $\mathfrak{g}$ on itself defined by the differential of Ad at the identity: $\mathrm{ad} \equiv (\mathrm{dAd})|_e : \mathfrak{g} \to gl(\mathfrak{g})$. It can easily be shown that for a given $A \in \mathfrak{g}$, A_{ad} is just the Lie derivative $A_{\mathrm{ad}}(B) = L_A(B) = [A, B]$.

The **trivial representation** maps all of G to the identity on a one-dimensional vector space; this representation is irreducible, and the corresponding Lie algebra representation maps all of $\mathfrak{g}$ to 0.

Closely related to linear representations are **projective representa-**

tions and **affine representations**, homomorphisms from G to $\text{Aut}(X)$ where X is a projective or affine space. Recall that a projective space is obtained from a vector space by taking all lines through the origin, i.e. by identifying scalar multiples of vectors. We can then view a projective representation as mapping each group element to an automorphism of a vector space V "ignoring vector length." An **affine space** is defined as a set on which a vector space acts freely and transitively as an additive group. Thus any two points in an affine space can be identified with the vector whose action relates them; i.e. an affine space "ignores the origin," so that vectors are defined between any two points, even if one is not the origin. In particular, a group representation that maps each group element to an automorphism on V that is an **affine map** (AKA inhomogeneous transformation), the sum of a linear and constant map, is an affine representation.

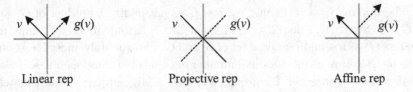

Linear rep   Projective rep   Affine rep

Figure 7.4.4 A linear rep maps each group element to a linear transformation of a vector; a projective rep maps each element to a linear transformation of a line; and an affine rep maps each element to a linear transformation of a vector plus an offset from the origin.

7.5   Classification of Lie groups

The classification of Lie groups and Lie algebras is a topic that is helpful, but not required, in understanding most of theoretical physics. Nevertheless, many of the concepts and terminology from this area are frequently seen in the physics literature, and so are covered here.

We have already learned two important facts with regard to the classification of Lie groups; namely, that every Lie group has connected components diffeomorphic to the normal subgroup of the identity component, and every connected Lie group G has a simply connected universal covering group G^*. G is then obtained from G^* by taking the quotient G^*/N, where N is a discrete normal subgroup that is isomorphic to $\pi_1(G)$. This discrete subgroup lies in the center of G^*. We then have a picture of how general

Lie groups are related to simply connected Lie groups, which are one-to-one with Lie algebras.

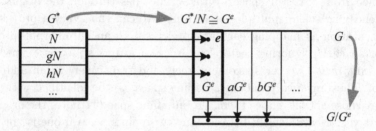

Figure 7.5.1 The identity component G^e is the quotient of the universal covering group G^* by a discrete normal subgroup N. A general Lie group G has G^e as a normal Lie subgroup.

G/G^e is called the **component group** of G, and is not in general a subgroup of G, so we cannot express G as a semidirect product of G^e and G/G^e. Similarly, since G^e is not in general a subgroup of G^*, we cannot express G^* as a semidirect product of N and G^e. Fortunately, most Lie groups in physics have at most a small number of connected components, so a classification of connected Lie groups will have a large impact. Unfortunately, classifying connected Lie groups is a vast and complicated field of study. Fortunately, in physics we can profitably narrow our focus to compact connected Lie groups, and still obtain important results. Unfortunately, even with this reduced scope, the subject is quite complicated.

7.5.1 *Compact Lie groups*

Compact Lie groups have many properties that make them an easier class to work with in general, including:

- Every compact Lie group has a faithful representation (and so can be viewed as a matrix group with real or complex entries)
- Every representation of a compact Lie group is similar to a unitary representation (and so is similar to a subgroup of $U(n)$ for some n)
- The Haar measure is bi-invariant on a compact Lie group

In terms of classification, it turns out that any compact Lie group has a Lie algebra that can be expressed as a direct sum of certain classes of Lie algebras. Recall that an abelian Lie algebra is one whose Lie bracket is identically zero. As one would expect, an abelian Lie group, one with an abelian Lie algebra, is also abelian in the group sense. A **simple Lie algebra** is a non-abelian one that has no nonzero proper ideal, and a **simple Lie group** is one with a simple Lie algebra. Sometimes a simple Lie group is defined to be one that is simple in a group sense. Note that this is a distinct definition, since in general a simple Lie group G in our sense is not simple in the group sense: it may have discrete normal subgroups corresponding to other Lie groups covered by G.

A Lie algebra that is a direct sum of simple Lie algebras is called a **semisimple Lie algebra**, and a **semisimple Lie group** is one with a semisimple Lie algebra. For any element A in a semisimple Lie algebra, there are elements B and C with $A = [B, C]$; thus it is a rough "opposite" to an abelian Lie algebra. Finally, a Lie algebra that is the direct sum of simple Lie algebras and abelian ones is called a **reductive Lie algebra**, and a **reductive Lie group** is one with a reductive Lie algebra (however, other definitions of reductive Lie group are sometimes used).

For our purposes, the important fact is this: any compact connected Lie group is a reductive group. Since every abelian connected compact Lie group is a direct product of copies of $U(1)$, and since every simple compact connected Lie group is a quotient of its universal covering group by a discrete normal subgroup, we can thus reduce the classification of compact connected Lie groups to that of simply connected compact simple Lie groups. This rather overwhelming scheme may possibly be clarified if viewed pictorially.

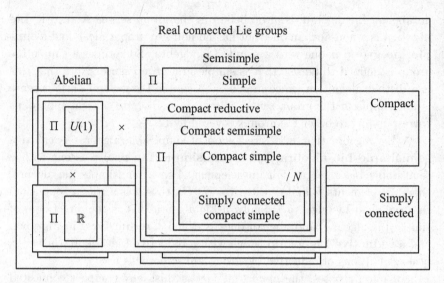

Figure 7.5.2 Any compact connected Lie group is of the form G/N, where G is a direct product of copies of $U(1)$ and simply connected compact simple Lie groups, and N is a normal discrete subgroup of G.

7.5.2 *Simple Lie algebras*

Our task is then to classify the simply connected compact simple Lie groups. Fortunately, the complex simple Lie algebras have been completely classified. Unfortunately, a complex Lie algebra $\mathfrak{g}$ has several distinct **real forms**, i.e. real Lie algebras whose complexifications are $\mathfrak{g}$. Fortunately, there is always a unique (up to isomorphism) **compact real form** of $\mathfrak{g}$, which is the only one that corresponds to a compact simple Lie group. It turns out that the universal covering group of any compact connected semisimple group is compact, so that in particular, each complex simple Lie algebra corresponds to a unique real simply connected compact simple Lie group.

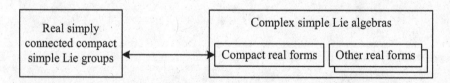

Figure 7.5.3 Classifying the compact connected Lie groups can be reduced to classifying the simply connected compact simple Lie groups, which are one-to-one with the compact real forms of the complex simple Lie algebras.

The complex simple Lie algebras consist of five **exceptional Lie algebras** g_2, f_4, e_6, e_7, and e_8, and four infinite series as follows.

Table 7.5.1 The infinite series of complex simple Lie algebras.

Series	Complex algebra	Compact real form	Other real form example
a_n $(n \geq 1)$	$sl\,(n+1, \mathbb{C})$	$su\,(n+1)$	$su\,(n+1-s, s)$
b_n $(n \geq 2)$	$so\,(2n+1, \mathbb{C})$	$so\,(2n+1)$	$so\,(2n+1-s, s)$
c_n $(n \geq 3)$	$sp\,(2n, \mathbb{C})$	$sp\,(n)$	$sp\,(2n, \mathbb{R})$
d_n $(n \geq 4)$	$so\,(2n, \mathbb{C})$	$so\,(2n)$	$so\,(2n-s, s)$

Notes: The last column gives one example of a non-compact real form; others exist as well. It is important to remember that we are complexifying real forms as manifolds, so that e.g. the $n \times n$ quaternionic matrix $sp(n)$ as a real manifold complexifies to $sp(2n, \mathbb{C})$, which also has the real form $sp(2n, \mathbb{R})$.

The series start at the indicated value of n to avoid duplicates in the list due to the following isomorphisms:

- $a_1 \cong b_1 \cong c_1$: for example $su(2) \cong so(3) \cong sp(1)$
- $b_2 \cong c_2$: for example $so(5) \cong sp(2)$
- $d_2 \cong a_1 \oplus a_1$: for example $so(4) \cong so(3) \oplus so(3)$
- $d_3 \cong a_3$: for example $so(6) \cong su(4)$

The third relation also leads to an isomorphism important in physics, $so(3, 1) \cong sl(2, \mathbb{C})$. It is interesting to note that the series of simple Lie algebras roughly correspond to "rotations" in $\mathbb{R}^n$, $\mathbb{C}^n$, and $\mathbb{H}^n$; in fact, it turns out this idea can be extended to the exceptional Lie algebras, which can be related to transformations on $\mathbb{O}$. The derivation of this classification uses geometric objects in $\mathbb{R}^n$ that we will not discuss called **root systems** and associated combinatorial objects called **Dynkin diagrams**.

Completing our task, the compact real forms determine the simply connected compact simple Lie groups to be $SU(n)$ (for $n \geq 2$), $Sp(n)$ (for $n \geq 3$), Spin(n) (for $n = 5$ or $n \geq 7$), and the simply connected compact exceptional Lie groups G_2, F_4, E_6, E_7, and E_8. Here Spin(n) is the universal covering group of $SO(n)$ for $n > 2$, to be discussed in Section 8.2.2. Thus our final classification is:

> Every compact connected Lie group is of the form G/N, where G is a direct product of $U(1)$ and the above simple Lie groups, and N is a normal discrete subgroup of G.

Regarding other types of Lie groups, we can note that above we have also classified the simple complex Lie algebras and groups. One can also show that any connected Lie group is topologically the product of a compact Lie group and a Euclidean space $\mathbb{R}^n$. We will end our discussion here, but we note that general Lie group theory provides various other statements concerning classifications.

7.5.3 *Classifying representations*

The irreducible representations of the complex simple Lie algebras have been fully classified, and as it turns out, these representations apply directly to their compact real forms. Using similar reasoning to above, the irreducible representations of the compact connected Lie algebras and Lie groups have also been fully classified.

An important example is that of the Lie algebra $su(2)$: up to similarity transformations there is one unique complex irreducible representation of $su(2)$ with dimension m for every $m \geq 1$. These representations are associated with angular momentum in quantum physics.

One can also classify the representations of finite groups, which are always completely reducible. In particular, **Young tableaux** are combinatorial diagrams used to enumerate the representations of the symmetric group S_n.

Chapter 8

Clifford groups

In this chapter we will go into some more detail on the structure of Clifford algebras, and then construct Lie groups within these algebras that are closely related to rotations and the concept of spin.

8.1 Classification of Clifford algebras

Like the classification of Lie groups, the classification of Clifford algebras is a topic that is helpful, but not required, in understanding most of theoretical physics. However, Clifford algebras and related constructs such as spinors are central to many modern physical theories, and so are worth exploring in detail.

Recall that the Clifford algebra over a given n-dimensional real vector space V with a pseudo inner product is defined to be the tensor algebra modulo the identification $vv \equiv \langle v, v \rangle$. The isomorphism classes of such Clifford algebras are then determined by the signature of the associated inner product, which we denote $C(r, s)$.

$\triangle$ Notation for Clifford algebras varies widely; in particular, r and s in our above notation are sometimes reversed, and $C(n)$ sometimes refers to either $C(0, n)$ or $C(n, 0)$.

8.1.1 *Isomorphisms*

The Clifford algebra can be viewed as a $\mathbb{Z}_2$-graded algebra, in that it can be decomposed into a direct sum of two vector subspaces generated by k-vectors with k either even or odd. The even subspace, denoted $C_0(r, s)$, is also a subalgebra, since the Clifford multiplication of two even k-vectors remains even.

By choosing $\hat{e}_0^2 = -1 \in C(r,s)$ and considering the algebra generated by the orthonormal basis $\hat{e}_0\hat{e}_i$ ($i \neq 0$), it is not hard to show that

$$C_0(r,s) \cong C(r, s-1).$$

Then the relationship $C_0(r,s) \cong C_0(s,r)$ leads to the isomorphism

$$C(r, s-1) \cong C(s, r-1).$$

One can also show that:

- $C(r,s) \otimes C(2,0) \cong C(r,s) \otimes \mathbb{R}(2) \cong C(s+2, r) \cong C(r+1, s+1)$
- $C(r,s) \otimes C(0,2) \cong C(r,s) \otimes \mathbb{H} \cong C(s, r+2)$
- $C(r,s) \otimes C(0,4) \cong C(r,s) \otimes \mathbb{H}(2) \cong C(r, s+4)$
- $C(r-4, s+4) \cong C(r,s)$
- The **periodicity theorem** (related to and sometimes referred to as **Bott periodicity**): $C(r+8, s) \cong C(r, s+8) \cong C(r,s) \otimes \mathbb{R}(16)$

The first isomorphism $C(r+1, s+1) \cong C(r,s) \otimes \mathbb{R}(2)$ means that we need only consider classifying Clifford algebras based on the values of $r - s$, and the periodicity theorem means that we can focus on values of $r - s$ mod 8.

In physics, the most important signatures are Euclidean and Lorentzian; specific isomorphisms for some of these Clifford algebras are listed in the following table. Note that since the first column covers all values of $r - s$ mod 8, it can be used to easily determine any other Clifford algebra.

Table 8.1.1 Isomorphisms for Clifford algebras of Euclidean and Lorentzian signatures.

n	$C(n,0) \cong C(1, n-1)$ $\cong C_0(n,1)$ $\cong C_0(1,n)$	$C(0,n) \cong C_0(n+1, 0)$ $\cong C_0(0, n+1)$	$C(n-1, 1)$
1	$\mathbb{R} \oplus \mathbb{R}$	$\mathbb{C}$	$\mathbb{C}$
2	$\mathbb{R}(2)$	$\mathbb{H}$	$\mathbb{R}(2)$
3	$\mathbb{C}(2)$	$\mathbb{H} \oplus \mathbb{H}$	$\mathbb{R}(2) \oplus \mathbb{R}(2)$
4	$\mathbb{H}(2)$	$\mathbb{H}(2)$	$\mathbb{R}(4)$
5	$\mathbb{H}(2) \oplus \mathbb{H}(2)$	$\mathbb{C}(4)$	$\mathbb{C}(4)$
6	$\mathbb{H}(4)$	$\mathbb{R}(8)$	$\mathbb{H}(4)$
7	$\mathbb{C}(8)$	$\mathbb{R}(8) \oplus \mathbb{R}(8)$	$\mathbb{H}(4) \oplus \mathbb{H}(4)$
8	$\mathbb{R}(16)$	$\mathbb{R}(16)$	$\mathbb{H}(8)$

Notes: Clifford multiplication corresponds to matrix multiplication in the isomorphic matrix algebra. Recall that our notation denotes e.g. the algebra of 2×2 matrices of quaternions as $\mathbb{H}(2)$.

We can also form the complexified version of the Clifford algebra $C(r, s)$, which is equivalent to considering the Clifford algebra generated by an inner product space $\mathbb{C}^n$. Since the signature is irrelevant in this case, we simply write $C^{\mathbb{C}}(n)$ where $r + s = n$. The complex Clifford algebras can be completely described by the following isomorphisms:

- $C^{\mathbb{C}}(2n) \cong \mathbb{C}(2^n)$
- $C^{\mathbb{C}}(2n + 1) \cong \mathbb{C}(2^n) \oplus \mathbb{C}(2^n)$

Note that this yields an isomorphism $C^{\mathbb{C}}(2n) \cong C(n, n{+}1) \cong C(n{+}2, n{-}1)$; in contrast, $C^{\mathbb{C}}(2n + 1)$ is not isomorphic to any real Clifford algebra. Also note that $C_0{}^{\mathbb{C}}(n) \cong C^{\mathbb{C}}(n - 1)$.

△ Although the above are all valid algebra isomorphisms, the original formulation of a Clifford algebra includes an extra structure: the generating vector space $\mathbb{R}^n$ that is explicitly embedded in $C(r, s)$. This extra structure is lost in these isomorphisms, since the choice of such an embedding is not in general unique.

8.1.2 *Representations and spinors*

With the matrix isomorphisms of the previous section in hand, the representation theory of Clifford algebras is quite simple, although the terminology is less so due to historical artifacts. First we note that for $\mathbb{K} \equiv \mathbb{R}$, $\mathbb{C}$, or $\mathbb{H}$, we have $\mathbb{K}(n) \cong \mathbb{K} \otimes \mathbb{R}(n)$, and as we will see in Section 8.1.3 on Pauli matrices, $\mathbb{C}$ is isomorphic to a subalgebra of $\mathbb{R}(2)$ and $\mathbb{H}$ is isomorphic to a subalgebra of $\mathbb{R}(4)$. This gives a representation of $\mathbb{C}(n)$ on $\mathbb{R}^{2n}$, which is in fact its only real faithful irreducible rep, and similarly gives $\mathbb{H}(n)$ a real faithful irrep on $\mathbb{R}^{4n}$, which is also unique. The unique real faithful irrep of $\mathbb{R}(n)$ is the obvious one on $\mathbb{R}^n$, so all real faithful representations of $\mathbb{K}(n)$ are direct products of the above irreducible ones.

Since all Clifford algebras are isomorphic to a matrix algebra of the form $\mathbb{K}(n)$ or $\mathbb{K}(n) \oplus \mathbb{K}(n)$, every Clifford algebra has either one unique real faithful irrep, called the **pinor rep**, or two real faithful irreps on the same vector space, called the **positive pinor rep** and the **negative pinor rep**. Similarly, every complexified Clifford algebra has either one or two complex faithful irreps, also called pinor reps, and from Section 7.4.4 we know that these are also faithful irreps of the real Clifford algebra on the same complex vector space. Various sub-classes of these faithful irreps have corresponding names that are used in physics:

- **Dirac rep**: for even n, the irrep of $C(r,s)$ on $\mathbb{C}^{2^{n/2}}$ given by the isomorphism $C^{\mathbb{C}}(n) \cong \mathbb{C}\left(2^{n/2}\right)$

- For odd n, there are two irreps of $C(r,s)$ on $\mathbb{C}^{2^{(n-1)/2}}$ given by the isomorphism $C^{\mathbb{C}}(n) \cong \mathbb{C}\left(2^{(n-1)/2}\right) \oplus \mathbb{C}\left(2^{(n-1)/2}\right)$; these are also sometimes called Dirac reps

- **Majorana rep**: for $r - s = 0$ or $2 \bmod 8$, the irrep of $C(r,s)$ on $\mathbb{R}^{2^{n/2}}$ given by the isomorphism $C(r,s) \cong \mathbb{R}\left(2^{n/2}\right)$

- For $r - s = 1 \bmod 8$, there are two irreps of $C(r,s)$ on $\mathbb{R}^{2^{(n-1)/2}}$ given by the isomorphism $C(r,s)^{\mathbb{C}} \cong \mathbb{R}\left(2^{(n-1)/2}\right) \oplus \mathbb{R}\left(2^{(n-1)/2}\right)$; these are also sometimes called Majorana reps

The even subalgebra $C_0(r,s) \cong C(r, s-1)$ thus also has either a unique faithful irrep, called the **spinor rep**, or two faithful irreps on the same vector space, called the right-handed and left-handed **chiral spinor reps** (AKA reduced spinor, semi-spinor, or half-spinor reps). The above subclasses of irreps then have names when they are applied to $C_0(r,s)$:

- **Weyl rep**: for even n, the two irreps of $C_0(r,s)$ on $\mathbb{C}^{2^{(n-2)/2}}$ given by the isomorphism $C_0^{\mathbb{C}}(n) \cong C^{\mathbb{C}}(n-1) \cong \mathbb{C}\left(2^{(n-2)/2}\right) \oplus \mathbb{C}\left(2^{(n-2)/2}\right)$

- **Majorana-Weyl rep**: for $r - s = 0 \bmod 8$, the two irreps of $C_0(r,s)$ on $\mathbb{R}^{2^{(n-2)/2}}$ given by the isomorphism $C_0(r,s) \cong C(s, r-1) \cong \mathbb{R}\left(2^{(n-2)/2}\right) \oplus \mathbb{R}\left(2^{(n-2)/2}\right)$

Thus a pinor rep may be irreducible as a representation of the Clifford algebra, but reducible when restricting the action to the even subalgebra, decomposing into two chiral spinor irreps. Elements of the vector space acted on by either pinor or spinor reps are called **spinors**, and are prefixed by the rep name (e.g. Dirac spinor); chiral spinors are sometimes called semi-spinors or half-spinors, and the vector space of spinors is called the **spinor space**.

In summary, since all faithful irreps of Clifford algebras are directly related to their isomorphisms in terms of algebras $\mathbb{K}(n)$, we can use these matrix algebras to illustrate the above classification scheme.

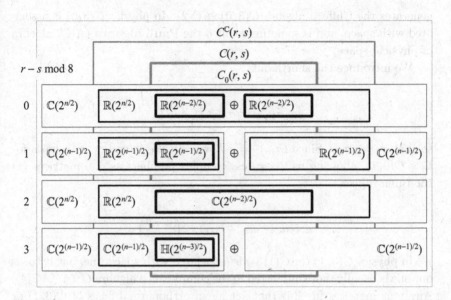

Figure 8.1.1 Classification of faithful Clifford algebra irreps using matrix algebra isomorphisms. For $r - s$ even ($\Rightarrow$ n even), $C^{\mathbb{C}}(n)$ has a Dirac rep, and $C_0{}^{\mathbb{C}}(n) \cong C^{\mathbb{C}}(n-1)$ splits into two Weyl reps. For $r - s = 0$ or 2 mod 8, $C(r,s)$ has a real Majorana rep, and in the first case $C_0(r,s)$ splits into two Majorana-Weyl reps. The next four lines are obtained by using $C(r,s) \otimes \mathbb{H}(2) \cong C(r, s+4)$, so there are no further isomorphisms to $gl(\mathbb{R})$ and thus no further way to generalize Majorana reps.

8.1.3 Pauli and Dirac matrices

The matrix isomorphisms of Clifford algebras are often expressed in terms of Pauli matrices. We will follow the common convention of using $\{i, j, k\}$ to represent matrix indices that are an even permutation of $\{1, 2, 3\}$; i also represents the square root of negative one, but the distinction should be clear from context.

The **Pauli matrices**

$$\sigma_1 \equiv \begin{pmatrix} 0 & 1 \\ 1 & 0 \end{pmatrix} \quad \sigma_2 \equiv \begin{pmatrix} 0 & -i \\ i & 0 \end{pmatrix} \quad \sigma_3 \equiv \begin{pmatrix} 1 & 0 \\ 0 & -1 \end{pmatrix}$$

are traceless, hermitian, determinant -1 matrices that satisfy the relations $\sigma_i \sigma_j = i\sigma_k$ and $\sigma_i \sigma_j \sigma_k = i$. They also all anti-commute and square to the identity $\sigma_0 \equiv I$; therefore, if we take matrix multiplication as Clifford multiplication, they act as an orthonormal basis of the vector space that generates the Clifford algebra $C(3,0) \cong \mathbb{C}(2)$. In physics $C(3,0)$ is associated with space, and is sometimes called the **Pauli algebra** (AKA algebra of physical space).

We introduce the shorthand

$$\sigma_{13} \equiv \sigma_1 \sigma_3 = \begin{pmatrix} 0 & -1 \\ 1 & 0 \end{pmatrix}$$

so that $\sigma_2 = i\sigma_{13}$. Since $(\sigma_{13})^2 = -I$, we can use it and σ_0 as a basis for $\mathbb{C} \cong C(0,1)$, allowing us to express complex numbers as real matrices via the isomorphism

$$a + ib \leftrightarrow a\sigma_0 + b\sigma_{13} = \begin{pmatrix} a & -b \\ b & a \end{pmatrix}.$$

In physics $C(3,1)$ (or $C(1,3)$) is associated with spacetime, but it turns out one is usually more interested in the complexified algebra $C^{\mathbb{C}}(4) \cong \mathbb{C}(4)$. Any four matrices in $\mathbb{C}(4)$ that act as an orthonormal basis of $C(3,1)$ or $C(1,3)$ to generate the Clifford algebra $C^{\mathbb{C}}(4)$ are called **Dirac matrices** (AKA gamma matrices), and denoted γ^i. A fifth related matrix is usually defined as $\gamma_5 \equiv i\gamma^0\gamma^1\gamma^2\gamma^3$. Many choices of Dirac matrices are in common use, a particular one being labeled the **Dirac basis** (AKA Dirac representation, standard basis). This is traditionally realized as a basis for $C(1,3)$:

$$\gamma^0 = \begin{pmatrix} I & 0 \\ 0 & -I \end{pmatrix}, \ \gamma^i = \begin{pmatrix} 0 & \sigma_i \\ -\sigma_i & 0 \end{pmatrix} \Rightarrow \gamma_5 = \begin{pmatrix} 0 & I \\ I & 0 \end{pmatrix}$$

Another common class of Dirac matrices requires γ_5 to be diagonal; this is called a **chiral basis** (AKA Weyl basis or chiral / Weyl representation). The meaning of γ_5 and "chiral" will be explained in the next section. A chiral basis for $C(1,3)$ is

$$\gamma^0 = \begin{pmatrix} 0 & I \\ I & 0 \end{pmatrix}, \ \gamma^i = \begin{pmatrix} 0 & \sigma_i \\ -\sigma_i & 0 \end{pmatrix} \Rightarrow \gamma_5 = \begin{pmatrix} -I & 0 \\ 0 & I \end{pmatrix},$$

and a chiral basis for $C(3,1)$ is

$$\gamma^0 = \begin{pmatrix} 0 & I \\ -I & 0 \end{pmatrix}, \ \gamma^i = \begin{pmatrix} 0 & \sigma_i \\ \sigma_i & 0 \end{pmatrix} \Rightarrow \gamma_5 = \begin{pmatrix} -I & 0 \\ 0 & I \end{pmatrix}.$$

Finally, a **Majorana basis** generates the Majorana rep $C(3,1) \cong \mathbb{R}(4)$. We can find such a basis by applying the previous isomorphism for complex numbers as real matrices to the Pauli matrices themselves, obtaining anti-commuting matrices in $\mathbb{R}(4)$ that square to the identity; if we then include an initial anti-commuting matrix that squares to $-I$, we get:

$$\gamma^0 = \begin{pmatrix} \sigma_{13} & 0 \\ 0 & -\sigma_{13} \end{pmatrix}, \; \gamma^1 = \begin{pmatrix} \sigma_1 & 0 \\ 0 & \sigma_1 \end{pmatrix}, \; \gamma^2 = \begin{pmatrix} 0 & -\sigma_{13} \\ \sigma_{13} & 0 \end{pmatrix},$$

$$\gamma^3 = \begin{pmatrix} \sigma_3 & 0 \\ 0 & \sigma_3 \end{pmatrix} \Rightarrow \gamma_5 = \begin{pmatrix} 0 & -\sigma_2 \\ -\sigma_2 & 0 \end{pmatrix}.$$

Note that all the above matrices representing positive signature basis vectors are Hermitian, and those representing negative signature basis vectors are anti-Hermitian; this is sometimes included as part of the defining requirements for Dirac matrices. Dirac or gamma matrices can also be generalized to other dimensions and signatures; in this light the Pauli matrices are gamma matrices for $C(3,0)$. If the dimension is greater than 5, γ_5 can be confused with γ^5; this is made worse by the fact that one can also define the **covariant Dirac matrices** $\gamma_i \equiv \eta_{ij}\gamma^j$.

$\triangle$ The Dirac matrices and γ_5 are defined in various ways by different authors. Most differ from the above only by a factor of ± 1 or $\pm i$; however, there is not much standardization in this area. Sometimes the Clifford algebra definition itself is changed by a sign; in this case the matrices represent a basis with the wrong signature, and according to our definition are not Dirac matrices. This is sometimes done for example when working with Majorana spinors, which only exist in $C(3,1)$ spacetime, yet where an author works nevertheless in the $C(1,3)$ "mostly minuses" signature.

$\triangle$ It is important to remember that the Dirac matrices are matrix representations of an orthonormal basis of the underlying vector space used to generate a Clifford algebra. So the Dirac and chiral bases are different representations of the orthonormal basis which generates the same matrix representation of $C^{\mathbb{C}}(4) \cong \mathbb{C}(4)$, which acts on vectors (spinors) in $\mathbb{C}^4$.

The standard basis for the quaternions $\mathbb{H} \cong C(0,2)$ can be obtained in terms of Pauli matrices via the association $\{1, i, j, k\} \leftrightarrow$

$\{\sigma_0, -i\sigma_1, -i\sigma_2, -i\sigma_3\}$. Thus a quaternion can be expressed as a complex matrix via the isomorphism

$$a + ib + jc + kd \leftrightarrow \begin{pmatrix} a - id & -c - ib \\ c - ib & a + id \end{pmatrix},$$

and composing this with the previous isomorphism for complex numbers as real matrices allows the quaternions to be expressed as a subalgebra of $\mathbb{R}(4)$.

The Pauli matrices also form a basis for the vector space of traceless hermitian 2×2 matrices, which means that $i\sigma_i$ is a basis for the vector space of traceless anti-hermitian matrices $so(3) \cong su(2)$. Thus any element of $SU(2)$ can be written $\exp\left(ia^j\sigma_j\right)$ for real numbers a^j. A similar construction is the eight **Gell-Mann matrices**, which form a basis for the vector space of traceless hermitian 3×3 matrices and so multiplied by i form a basis for $su(3)$.

$\triangle$ Since the Pauli matrices have so many potential roles, it is important to understand what use a particular author is making of them.

8.1.4 *Chiral decomposition*

As we have seen, the Dirac and Majorana reps are in fact isomorphisms from $C^{\mathbb{C}}(r, s)$ and $C(r, s)$, and so are faithful and irreducible; however, as we have also seen they are sometimes reducible when restricting their action to the even subalgebra, decomposing into two chiral irreps. These chiral spinor reps of $C_0(r, s)$ can be obtained by projection, revealing some important attributes.

For $A \in C_0(r, s)$ and even n, one can verify that the operators

$$P_{\pm}(A) \equiv \frac{1}{2}\left(A \pm \sqrt{\Omega^2}\,\Omega A\right)$$

are **orthogonal projections**, i.e. they are idempotent, $P_{\pm}^2 = P_{\pm}$, with $P_{\pm}P_{\mp} = 0$ and $P_+ + P_- = 1$. Since the unit n-vector Ω of $C(r, s)$ commutes with any $A \in C_0(r, s)$, we then have the decomposition

$$C_0(r, s) \cong P_+(C_0(r, s)) \oplus P_-(C_0(r, s)).$$

Note that this decomposition is not possible if n is odd, since then $\Omega A \notin C_0(r, s)$ and so $P_{\pm}(C_0(r, s)) \notin C_0(r, s)$. For even n, the quantity $\Omega^2 =$

$(-1)^{n(n-1)/2+s}$ must be positive in order to obtain a real square root. This is only true if $r - s = 0$ or 4 mod 8, and as we saw previously, only in the first case is the resulting algebra isomorphic to a real matrix algebra and thus a Majorana-Weyl rep. This restriction is avoided if we apply the decomposition to $C_0{}^{\mathbb{C}}(n)$, which is why a Weyl rep exists for any even n.

In the present context $\gamma_5 \equiv \sqrt{\Omega^2}\,\Omega$ is sometimes called the **chirality operator**, and is the generalization of γ_5 to arbitrary signature and dimension. This explains the name of the "chiral basis" Dirac matrices from Section 8.1.3, since they diagonalize the chirality operator. The specific chiral bases we listed allow us write the Dirac spinor ψ as stacked Weyl spinors, since

$$P_+ (\psi) = \frac{1}{2} (I + \gamma_5)\, \psi = \begin{pmatrix} 0 & 0 \\ 0 & I \end{pmatrix} \begin{pmatrix} \psi_L \\ \psi_R \end{pmatrix} = \begin{pmatrix} 0 \\ \psi_R \end{pmatrix},$$

where P_- similarly projects to the ψ_L half-spinor. ψ_R along with the associated rep and projection are called "right-handed" due to the fact that in physics they correspond to particles "spinning" around an axis aligned with the particle's momentum, using the right-hand rule.

In the case of a Lorentzian signature, we can also consider **time reversal** and **parity** operators, which reverse the sign of either the negative signature basis vector or the $(n - 1)$ positive signature basis vectors. In either case, for even n this consists of reversing an odd number of basis vectors, so that $\Omega \to -\Omega$, and thus under either operation the chiral spinor reps are swapped: $P_\pm (C_0 (r, s)) \to P_\mp (C_0 (r, s))$.

8.2 Clifford groups and representations

Rotations on the vector space V, i.e. linear transformations that preserve the inner product, can be expressed in terms of Clifford multiplication, regardless of the signature. This allows us to link spacetime transformations to spinor transformations.

8.2.1 *Reflections*

In the Clifford algebra, any "unit" vector, i.e. a vector u with $\langle u, u \rangle = \pm 1$, has inverse $u^{-1} = u / \langle u, u \rangle = \pm u$. It turns out that for any vector v, the quantity $R_u(v) \equiv -uvu^{-1}$ is the reflection of v in the hyperplane orthogonal to the unit vector u.

We can see this by decomposing v into a part $v_\parallel = (\langle u, v \rangle / \langle u, u \rangle)\, u$ that is parallel to u and the remaining part $v_\perp$ that is orthogonal to u. Since parallel vectors commute, we have $-uv_\parallel u^{-1} = -v_\parallel$. In contrast, orthogonal vectors anti-commute, so that we have $-uv_\perp u^{-1} = v_\perp$, and thus

$$R_u(v) = -uvu^{-1} = -u\left(v_\perp + v_\parallel\right)u^{-1} = v_\perp - v_\parallel.$$

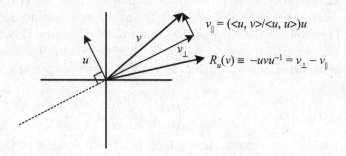

Figure 8.2.1 Any unit vector u can be used to reflect vectors across the hyperplane orthogonal to u.

8.2.2 *Rotations*

Any element of the orthogonal group $O(r, s)$ is a rotation and/or reflection, and a well-known result is that any such transformation can be obtained as a product of reflections. Thus every element of $O(r, s)$ corresponds to the Clifford product of some k unit vectors via

$$R_{u_k \cdots u_1}(v) \equiv R_{u_k}\left(\cdots\left(R_{u_1}(v)\right)\cdots\right) = (-1)^k u_k \cdots u_1 v u_1^{-1} \cdots u_k^{-1}.$$

The elements of $C(r, s)$ that have the form of a Clifford product of unit vectors $U = u_k \cdots u_1$ form a Lie group denoted $\text{Pin}(r, s)$ and called a **Clifford group** (AKA Pin group). In terms of the reverse operation of geometric algebra, the elements $U \in \text{Pin}(r, s)$ are those that satisfy $U\widetilde{U} = \pm 1$. R forms a homomorphism from $\text{Pin}(r, s)$ to $O(r, s)$ defined by $U \mapsto R_U$, where

$$R_U(v) = (-1)^k U v U^{-1}.$$

This homomorphism is two-to-one, since R_U and R_{-U} map to the same transformation, so $\text{Pin}(r, s)$ is a double covering of $O(r, s)$.

Now, elements of $SO(r, s)$ are pure rotations, i.e. they are obtained as a product of an even number of reflections. Therefore the **special Clifford**

group (AKA Spin group) $\text{Spin}(r,s) \equiv \text{Pin}(r,s) \cap C_0(r,s)$ is a double covering of $SO(r,s)$, using the restriction of R_U to even elements:

$$R_U^S(v) = UvU^{-1}$$

△ It is important to remember that rotations in 4 dimensions and higher do not follow many intuitive ideas from 3 dimensions. In particular, a rotation can have more than one plane of rotation (where rotated vectors in the plane stay in the plane), and therefore can require more than two reflections.

These relationships are depicted in the following diagram.

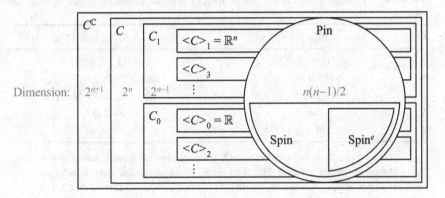

Figure 8.2.2 The Clifford group Pin and its even subgroup Spin are generated by the unit elements of the Clifford algebra C. C_1 and C_0 both have dimension 2^{n-1} as manifolds, and Pin, Spin, and Spin^e all have dimension $n(n-1)/2$.

△ Some potential sources of confusion can be avoided by remembering that $\text{Pin}(r,s)$ is the *group* generated by Clifford multiplication on the unit elements of the *algebra* $C(r,s)$. Thus the elements of $\text{Pin}(r,s)$ can only be multiplied with each other and always have inverses, while the elements of $C(r,s)$ can be multiplied by scalars and added, but may not have multiplicative inverses.

8.2.3   *Lie group properties*

Although the Lie groups Pin (r, s) and Spin (r, s) are double covers of $O(r, s)$ and $SO(r, s)$ (except for $r = s = 1$, see table below), they are in general not simply connected and so are not universal covering groups. However, it turns out that Spin $(n) \equiv$ Spin $(n, 0) \cong$ Spin $(0, n)$ is simply connected for $n > 2$, and thus is the universal covering group of $SO(n)$. In addition, for $n > 2$ the identity component Spin $(n, 1)^e \cong$ Spin $(1, n)^e$ is simply connected and is the double cover of $SO(1, n)^e \cong SO(n, 1)^e$. Spin (r, s) does not always have a simple description in terms of common matrix groups. Specific isomorphisms for the identity component of the first few Spin groups in Euclidean and Lorentzian signatures are listed in the following table.

Table 8.2.1 Isomorphisms for some Spin groups as matrices under multiplication.

n	Spin $(n, 0) \cong$ Spin $(0, n)$	Spin $(n - 1, 1)^e \cong$ Spin $(1, n - 1)^e$
2	$U(1) \cong SO(2)$	$SO(1, 1)^e \cong \mathbb{R}^+$
3	$Sp(1) \cong SU(2)$	$Sp(2, \mathbb{R}) \cong SL(2, \mathbb{R})$
4	$Sp(1) \oplus Sp(1) \cong SU(2) \oplus SU(2)$	$Sp(2, \mathbb{C}) \cong SL(2, \mathbb{C})$
5	$Sp(2)$	$Sp(1, 1)$
6	$SU(4)$	$SL(2, \mathbb{H})$

Notes: Spin(2) is isomorphic to $SO(2)$ as an abstract group, but is nevertheless a double cover of $SO(2)$ in terms of its action on vectors; however it is not simply connected. Spin$(1, 1)^e$ is not a double cover of $SO(1, 1)^e$, which is instead obtained by imposing the restriction $U\widetilde{U} = 1$ on Spin$(1, 1)$ to obtain $GL(1, \mathbb{R}) \cong \{R - 0\}$. Spin$(2, 1)^e$ is not simply connected, but is a double cover of $SO(2, 1)^e$. In all higher dimensions Spin$(n, 0)$ and Spin$(n, 1)^e$ are always the simply connected double covers of $SO(n)$ and $SO(n, 1)^e$. $Sp(1, 1)$ is defined in a straightforward way, but $SL(2, \mathbb{H})$ (not defined here) is more tricky due to non-commutativity.

Since Spin(r, s) is a cover of $SO(r, s)$, its Lie algebra is $so(r, s)$, which turns out to be the space of bivectors $\Lambda^2 \mathbb{R}^n$ under the Lie commutator using Clifford multiplication: $[A, B] \equiv AB - BA$. A more general construct sometimes seen is the Lie group generated by all invertible elements of $C(r, s)$, called the **Lipschitz group** (AKA Clifford group); its Lie algebra is $C(r, s)$ itself, also under the Lie commutator using Clifford multiplication.

We consequently have an equivalent definition of Spin $(r, s)^e$ as the group generated by the exponentials of bivectors under Clifford multiplication.

For the compact Lie groups $\mathrm{Spin}\,(n,0)$ $(n > 1)$, every element is then the exponential of a bivector; as it turns out, this is also true for $\mathrm{Spin}\,(n,1)^e$ for $n > 3$, and every element of $\mathrm{Spin}\,(3,1)^e$ is plus or minus such an exponential.

8.2.4 *Lorentz transformations*

In Euclidean space (of any dimension), we can write any element $U = \exp(B)$ of $\mathrm{Spin}(n,0)$ in terms of "unit" 2-blades $\hat{B}_i^2 = -1$ as $U = \exp\left(-\theta^i \hat{B}_i/2\right)$. Under the homomorphism R, it is not hard to show that this element then corresponds to active rotations by θ^i in the planes defined by the 2-blades $\hat{B}_i$.

In 2 dimensions, if $\hat{B} = \hat{e}_1\hat{e}_2 = \hat{e}_1 \wedge \hat{e}_2$, then

$$U = \exp\left(-\frac{\theta}{2}\hat{e}_1\hat{e}_2\right) = \cos\left(\frac{\theta}{2}\right) - \sin\left(\frac{\theta}{2}\right)\hat{e}_1\hat{e}_2,$$

and $R_U(v) = UvU^{-1}$ can be seen to be a rotation of v by θ. Thus this element can also be written as $U = u_1u_2$, the Clifford product of two unit vectors separated by an angle of $\theta/2$, which corresponds to the same rotation $R_U(v) = Uv\widetilde{U} = u_1u_2vu_2u_1$. Since $\hat{B}^2 = -1$, it naturally maps to i in the rep on $U(1)$, thus corresponding to a spinor "rotation" of $\theta/2$.

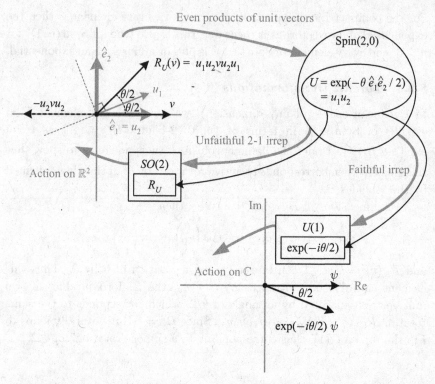

Figure 8.2.3 An element $U = \exp(-\theta\hat{e}_1\hat{e}_2/2) = u_1u_2$ of Spin(2,0) corresponds to a rotation by θ which can be written $R_U(v) = \exp(-\theta\hat{e}_1\hat{e}_2/2)v\exp(+\theta\hat{e}_1\hat{e}_2/2) = u_1u_2vu_2u_1$, where u_1 and u_2 are unit vectors separated by an angle of $\theta/2$. Since the "rotation" in $U(1)$ is by only $\theta/2$, a complete rotation in $\mathbb{R}^2$ corresponds to half of one in $\mathbb{C}$, i.e. for $\theta = 2\pi$, $\exp(-i2\pi/2) = -1$.

$\triangle$ In arbitrary dimension and signature, the idea that a spinor "rotation" by $\theta/2$ corresponds to a rotation by θ is less clear; but it is always true that a spinor sign reversal corresponds to a return to the original state (e.g. a rotation by 2π), since $U = -1$ corresponds to $R_U(v) = (-1)v(-1) = v$.

In 3 dimensions, we can choose an orthonormal basis $\hat{e}_i$ and write the element as

$$U = \exp\left(-\sum\frac{\theta^i}{2}\hat{e}_j\hat{e}_k\right),$$

where the unit vector $\hat{u} = \hat{u}^i \hat{e}_i$ is defined by $\theta \hat{u}^i \equiv \theta^i$, the sum is over all indices $\{i, j, k\}$ that are even permutations of $\{1, 2, 3\}$, and U can be seen to correspond to a rotation by θ in the plane perpendicular to $\hat{u}$. Note that we cannot split this into a product of exponentials, since these bivectors do not commute.

In spacetime, an element U of $\mathrm{Spin}(3, 1)^e$ corresponds to a restricted (i.e. proper orthochronous) Lorentz transformation, an element of $SO(3, 1)^e$. Choosing an orthonormal basis $\hat{e}_\mu$, a boost of rapidity ϕ in the direction of the unit vector $\hat{v} = \hat{v}^i \hat{e}_i$ corresponds to the element $\exp(-\phi^i \hat{e}_0 \hat{e}_i / 2)$ of $\mathrm{Spin}(3, 1)^e$, where $\phi^i \equiv \phi \hat{v}^i$. Similarly, a rotation by θ in the space-like plane perpendicular to $\hat{u} = \hat{u}^i \hat{e}_i$ corresponds to the element $\exp(-\theta^i \Omega \hat{e}_0 \hat{e}_i / 2)$, where $\theta^i \equiv \theta \hat{u}^i$. Any element of $\mathrm{Spin}(3, 1)^e$ can then be written

$$U = \pm \exp\left(-\frac{\phi^i}{2} \hat{e}_0 \hat{e}_i - \frac{\theta^i}{2} \Omega \hat{e}_0 \hat{e}_i \right).$$

Note that this decomposition into boost and rotation terms depends on the specific basis, and that again we cannot split this into a product of exponentials, since $\hat{e}_0 \hat{e}_i$ and $\hat{e}_0 \hat{e}_j$ do not commute.

If we start with a transformation U instead of a basis, we can then choose a basis that lets us split the exponential factors. We know that U is of the form $\pm \exp(B)$, where B is a bivector in $C(3, 1)$. It is not hard to show that any non-null bivector $B^2 \neq 0$ can be written in the form $B = -\phi \hat{B}/2 - \theta \Omega \hat{B}/2$, where $\hat{B}^2 = 1$. Since Ω commutes with any bivector and $\Omega^2 = -1$, it acts like the imaginary unit $\sqrt{-1}$, and the multiplication of exponentials follows the scalar rule. We can thus write U in the form

$$U = \exp\left(-\frac{\phi}{2} \hat{B} \right) \exp\left(-\frac{\theta}{2} \Omega \hat{B} \right).$$

Note that we have dropped the $\pm$ factor since $-\exp(i\theta) = \exp(i(\pi - \theta))$.

This decomposition means that U determines a class of orthonormal basis vectors for $\mathbb{R}^4$ in which $\hat{B} = \hat{e}_0 \hat{v}$, where $\hat{v}$ is a unit vector in space $\hat{e}_1 \hat{e}_2 \hat{e}_3$. In such a basis, the first term corresponds to a Lorentz boost of rapidity ϕ in the $\hat{v}$ direction, and the second term corresponds to a rotation by θ in the space-like plane perpendicular to $\hat{v}$. These orthogonal planes are invariant under the total transformation, i.e. vectors in these planes remain in them.

The above decomposition fails in the case $B^2 = 0$ because this corresponds to a "light-like" or "parabolic" Lorentz transformation, which leaves a single null plane (orthogonal to itself) invariant. This case is also the reason why the general form has to be prefixed by the $\pm$ factor.

> △ It is important to remember that the treatment here is for active ro-
> tations. Passive rotations (those that rotate the coordinate axes while
> leaving the vector fixed) are in the opposite direction, so to represent
> them we would multiply both angles by -1 in the above expressions.

8.2.5 Representations in spacetime

The pinor and spinor reps of $C(r,s)$ and $C_0(r,s)$ turn out to still be irreducible when their actions are restricted to $\text{Pin}(r,s)$ and $\text{Spin}(r,s)$, so
the representation theory of Clifford groups is identical to that of Clifford
algebras. In this section we will list concrete representations for the various quantities we have defined in the two most common vector spaces in
physics, corresponding to space and spacetime. We will choose bases that
highlight the isomorphism $C(3,0) \cong C_0(3,1)$ and use the Pauli matrices
σ_i.

Dealing first with space, we can view $\text{Spin}(3,0) \cong SU(2) \cong Sp(1)$ as
either a group of quaternions sitting inside $C_0(3,0) \cong \mathbb{H}$, or as a group of
complex matrices sitting inside $C(3,0) \cong \mathbb{C}(2)$. Both approaches can be
fruitful, but we will focus on the latter, which since $C^{\mathbb{C}}(3) \cong \mathbb{C}(2) \oplus \mathbb{C}(2)$
can be viewed as the "odd Dirac" rep for signature $(3,0)$.

We can represent a chosen orthonormal basis $\hat{e}_i$ of $\mathbb{R}^3$ by the matrices
σ_i, which have the correct properties of squaring to 1 and anti-commuting
under Clifford (matrix) multiplication. These basis vectors then generate
the Clifford algebra $C(3,0) \cong \mathbb{C}(2)$. The bivectors are thus naturally represented by $\sigma_i \sigma_j = i\sigma_k$, so that the elements of $\text{Spin}(3,0) \cong SU(2)$ are of
the form

$$\exp\left(-i\frac{\theta^i}{2}\sigma_i\right),$$

and $\hat{u} = \hat{u}^i \hat{e}_i$ is a unit length vector defined by $\theta \hat{u}^i \equiv \theta^i$. This corresponds
to a rotation by θ in the plane orthogonal to $\hat{u}$.

In spacetime, $\text{Spin}(3,1)^e \cong \text{Spin}(1,3)^e \cong SL(2,\mathbb{C})$ is a group of complex matrices sitting inside inside $C_0(3,1) \cong C_0(1,3) \cong \mathbb{C}(2)$, which can
be viewed as the Weyl rep, the chiral decomposition of the Dirac rep
$C^{\mathbb{C}}(4) \cong \mathbb{C}(4)$. We choose to represent the bivectors $\hat{e}_0\hat{e}_i$ by the matrices σ_i (which again have the correct properties of squaring to 1 and
anti-commuting), the remaining bivectors $\hat{e}_i\hat{e}_j = \hat{e}_0\hat{e}_i\hat{e}_0\hat{e}_j$ therefore being represented by the matrices $\sigma_i \sigma_j = i\sigma_k$. Since every element of

Spin $(3,1)^e \cong$ Spin $(1,3)^e$ is of the form of plus or minus the exponential of a bivector, every element can then be written as

$$U_L = \pm \exp\left(-i\frac{\theta^i}{2}\sigma_i - \frac{\phi^i}{2}\sigma_i\right),$$

where $\theta\hat{u}^i \equiv \theta^i$, $\phi\hat{v}^i \equiv \phi^i$, and the element corresponds to a rotation by θ in the space-like plane perpendicular to the unit vector $\hat{u} = \hat{u}^i\hat{e}_i$ and a Lorentz boost of rapidity ϕ in the direction of the unit vector $\hat{v} = \hat{v}^i\hat{e}_i$.

Note that we also have the choice to represent the bivectors $\hat{e}_0\hat{e}_i$ by the matrices $-\sigma_i$ (which also square to 1 and anti-commute), which leaves the remaining bivectors $\hat{e}_i\hat{e}_j$ still represented by $\sigma_i\sigma_j = i\sigma_k$, so that the alternative general bivector form is

$$U_R = \pm \exp\left(-i\frac{\theta^i}{2}\sigma_i + \frac{\phi^i}{2}\sigma_i\right).$$

These representations can be seen to be inequivalent, i.e. there is no similarity transformation that goes between them. Referring back to the Dirac matrices in the chiral basis for $C(3,1)$

$$\gamma^0 = \begin{pmatrix} 0 & I \\ -I & 0 \end{pmatrix}, \; \gamma^i = \begin{pmatrix} 0 & \sigma_i \\ \sigma_i & 0 \end{pmatrix} \Rightarrow \gamma^0\gamma^i\psi = \begin{pmatrix} \sigma_i & 0 \\ 0 & -\sigma_i \end{pmatrix}\begin{pmatrix} \psi_L \\ \psi_R \end{pmatrix} = \begin{pmatrix} \sigma_i\psi_L \\ -\sigma_i\psi_R \end{pmatrix},$$

we can see that the first representation of the bivectors $\hat{e}_0\hat{e}_i$ by the matrices σ_i corresponds to the left-handed Weyl rep, explaining the subscripts above.

$\triangle$ Note that with a $(1,3)$ signature, the bivector reps would remain the same, but referring back to the Dirac matrices in the chiral basis for $C(1,3)$

$$\gamma^0 = \begin{pmatrix} 0 & I \\ I & 0 \end{pmatrix}, \; \gamma^i = \begin{pmatrix} 0 & \sigma_i \\ -\sigma_i & 0 \end{pmatrix} \Rightarrow \gamma^0\gamma^i\psi = \begin{pmatrix} -\sigma_i\psi_L \\ \sigma_i\psi_R \end{pmatrix},$$

we would swap the definitions U_R and U_L above. In mapping the above to other treatments, it is important to remember that the stacking order of the column vector depends on the choice of Dirac matrices, but the form of the left and right Lorentz transformation reps depend only on the signature; e.g. we can get alternative Dirac matrices by multiplying γ^0 by -1, but this does the same to γ_5, thus swapping the stacking order of the Weyl spinors in chiral decomposition and leaving the forms of U_R and U_L the same. Also again note that many treatments are for passive rotations, which would multiply all angles by -1.

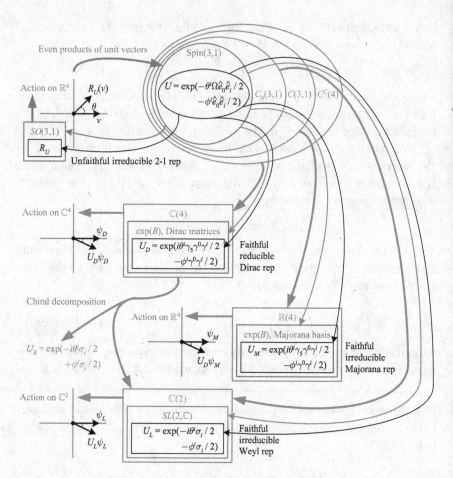

Figure 8.2.4 An element U of the group $Spin(3,1)$ acts as a Lorentz transformation on spacetime via the rep $SO(3,1)$, and as an exponential of bivectors expressed as multiplied pairs of Dirac matrices on spinors (in $\mathbb{C}^4$ for any basis, or in $\mathbb{R}^4$ for the Majorana basis). The rep acting on $\mathbb{C}^4$ is reducible to the two Weyl irreps, where U then acts as an exponential of bivectors expressed as Pauli matrices σ_i and $i\sigma_i$ on half-spinors. The bivector reps acting on spinors can be expressed in terms of the angles that generate Lorentz boosts and rotations on spacetime. We drop the $\pm$ before the exponentials, choosing the positive transformations which are "small" for both vectors and spinors.

8.2.6 *Spacetime and spinors in geometric algebra*

In geometric algebra, the **rotor group** is the Lie group obtained by restricting Spin (r, s) to elements whose reverse is their inverse, i.e. elements which satisfy $U\widetilde{U} = 1$. For $n > 2$ this restriction results in the identity component, i.e the rotor group is just Spin^e. Thus in this context for $n > 2$ we can write $R_U^e(v) = Uv\widetilde{U}$. This operator can also be applied to any multivector $A = \Sigma \langle A \rangle_k$ to yield the "rotated" multivector $UA\widetilde{U}$. Under this operation, each k-blade, consisting of the exterior product of k vectors, is replaced with the exterior product of k rotated vectors.

The representation of the isomorphism $C_0(1,3) \cong C(3,0)$ effected by $\hat{e}_0\hat{e}_i \to \sigma_i$ is sometimes called a **space-time split** in geometric algebra, since the resulting basis of $C(3,0)$ reflects (and depends upon) the particular chosen orthonormal basis $\hat{e}_i$ of $C_0(1,3)$. An event $x \in C(1,3)$ with spacetime coordinates $x^\mu\hat{e}_\mu$ is represented by $\hat{e}_0x = x^0 + x^i\sigma_i$ in $C(3,0)$; such a linear combination of scalar and vector in $C(3,0)$ is then called a **paravector** (although this term is sometimes used differently). This scheme can be used to treat relativistic physics in a condensed manner. Note that a space-time split "preserves" the scalar and pseudo-scalar basis: $I \to I$ and $\Omega \to \Omega$. If spacetime is instead represented by the "mostly pluses" signature algebra $C(3,1)$, $-x\hat{e}_0$ can be used as the space-time split in order to make the signs come out right. Note that the rep $\hat{e}_i\hat{e}_0 \to \sigma_i$ is sometimes used instead, reversing the order of the $\hat{e}_0$ multiplication used to get paravectors.

There is also an interesting alternative to the standard definition of Dirac and Weyl spacetime spinors (as vectors acted on by a faithful complex representation of Spin$(3,1)^e$), which instead considers these spinors as elements of the Clifford algebra associated with space. The Dirac spinors are vectors in $\mathbb{C}^4$, a complex vector space of dimension 4 that decomposes into two orthogonal 2-dimensional complex subspaces which are each invariant under the action of Spin$(3,1)^e$. Now, the even subalgebra $C_0(3,1) \cong C(3,0) \cong \mathbb{C}(2)$ can also be viewed as a complex vector space of dimension 4. The action of Spin$(3,1)^e$ on $C_0(3,1)$ by Clifford multiplication is linear, and $C_0(3,1)$ decomposes into two spaces invariant under Spin$(3,1)^e$: the bivectors that are real linear combinations of $e_0e_i \cong \sigma_i$ have negative determinant, while linear combinations of the remaining bivectors $e_ie_j \cong \sigma_i\sigma_j = i\sigma_k$ have positive determinant, as do the scalars. The pseudo-scalars are real multiples of $\Omega = e_0e_1e_2e_3$, and so have negative determinant. Thus an element of Spin$^e(3,1) \cong SL(2,\mathbb{C})$, having determinant $+1$, leaves invariant these positive and negative determinant subspaces under Clifford multiplication. Note that as $C(3,0)$, these subspaces are exactly the even and odd subspaces $C_0(3,0)$ and $C_1(3,0)$. From Section 8.2.5 we know that the Dirac and Weyl reps are the unique faithful reps of $\mathrm{Spin}(3,1)^e$ on $\mathbb{C}^4$ and $\mathbb{C}^2$, so the above two representations must be equivalent to these.

This alternative definition of spinors then readily generalizes to any dimension, i.e. spinors can be defined as elements of the 2^{n-1}-dimensional vector space $C_0(r,s) \cong C(r,s-1)$ acted on by $\mathrm{Spin}(r,s)^e$ via Clifford multiplication. However, it is important to note that despite the accidental equivalency in signature $(3,1)$, for other signatures and dimensions these definitions are quite distinct. In particular, the above decomposition of $C_0(3,1) \cong \mathbb{C}(2)$ as a vector space of spinors under the action of $\mathrm{Spin}(3,1)^e$ is completely unrelated to the chiral decomposition of the Dirac rep $C^{\mathbb{C}}(3,1) \cong \mathbb{C}(4)$ when restricted to $C_0(3,1)$. This is underscored by the fact that while the Dirac rep decomposes as a rep of any part of the even subalgebra, the decomposition of the spinor space $C_0(3,1)$ only occurs under the action of the identity component $\mathrm{Spin}(3,1)^e$: the determinant is not preserved under the action of $\mathrm{Spin}(3,1) \cong SL^{\pm}(2,\mathbb{C})$, and so there is no decomposition in this case. Lastly, note that this gives us a new characterization of $\mathrm{Spin}(3,1)^e$ as plus or minus the exponentials of the Lie algebra of vectors and bivectors in $C(3,0)$.

Chapter 9

Riemannian manifolds

In this chapter we introduce two additional structures on a differentiable manifold. First we consider the "parallel transport" of a vector, which allows a vector at one point on the manifold to be "transported" along a path to another point, where it can then be compared to other vectors at the new point. This idea gives rise to a number of interdependent quantities, and is particularly important in physics, where it is generalized to gauge theories.

We then consider the introduction of a metric, an inner product in each tangent space that permits us to define lengths of vectors and angles between them. A metric determines a unique associated parallel transport, and is the fundamental quantity in general relativity. We then touch upon some other structures on manifolds that appear in physics.

9.1 Introducing parallel transport of vectors

9.1.1 *Change of frame*

In this section we will introduce a number of quantities that depend on a choice of frame. We can then consider a **change of frame**, a linear transformation of the basis e_μ of each tangent space on the manifold. Since an arbitrary manifold is not necessarily parallelizable, a frame is already a local construct; therefore we assume a change of frame preserves orientation, and define it to be a tensor field $(\gamma^{-1})^\nu{}_\mu$ of elements in $GL(n, \mathbb{R})^e$ smoothly defined in the region $U \subset M$ where the frame is defined. We write γ^{-1} for the change of frame so that the components of a vector field w^μ transform according to $w'^\nu = \gamma^\nu{}_\mu w^\mu$.

$\triangle$ Note that the vector field w is an intrinsic object that is unaffected by a change of frame; it is only the components w^μ that transform. It is common to use γ to denote a matrix and then write $w' = \gamma w$ or $e' = e\gamma^{-1}$, where w is understood to be a column vector of components and e is understood to be a row matrix of basis vectors. We will attempt to always explicitly mention whether variables are matrices, and will in most cases show indices when referring to components to avoid confusion with the intrinsic vector field.

Below we summarize how some common objects transform under a change of frame $(\gamma^{-1})^\nu{}_\mu$.

Table 9.1.1 Transformations under a change of frame $(\gamma^{-1})^\nu{}_\mu$.

Construct	In the original frame	In the transformed frame
Frame	e_μ	$e'_\nu = (\gamma^{-1})^\mu{}_\nu e_\mu$
Dual frame	β^μ	$\beta'^\nu = \gamma^\nu{}_\mu \beta^\mu$
Vector field components	w^μ	$w'^\nu = \gamma^\nu{}_\mu w^\mu$
1-form components	φ_μ	$\varphi'_\nu = (\gamma^{-1})^\mu{}_\nu \varphi_\mu$
Linear transformation	$\Theta^\mu{}_\nu$	$\Theta'^\lambda{}_\sigma = \gamma^\lambda{}_\mu \Theta^\mu{}_\nu (\gamma^{-1})^\nu{}_\sigma$

9.1.2   *The parallel transporter*

By definition, for a vector w at a point p of an n-dimensional manifold M, **parallel transport** assigns a vector $\parallel_C (w)$ at another point q that is dependent upon a specific path C in M from p to q.

To see that this dependence upon the path matches our intuition, we can consider a vector transported in what we might consider to be a "parallel" fashion along the edges of an eighth of a sphere. In this example, the sphere is embedded in $\mathbb{R}^3$ and the concept of "parallel" corresponds to incremental vectors along the path having a projection onto the original tangent plane that is parallel to the original vector.

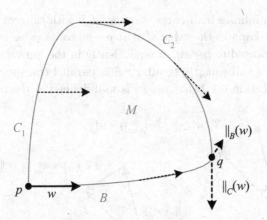

Figure 9.1.1 A vector w transported in what we intuitively consider to be a "parallel" way along two different paths (B and $C = C_1 + C_2$) on a surface results in two different vectors.

The **parallel transporter** is therefore a map $\|_C \colon T_pM \to T_qM$, where C is a curve in M from p to q. To match our intuition we also require that this map be linear (i.e. parallel transport is assumed to preserve the vector space structure of the tangent space); that it be the identity for vanishing C; that if $C = C_1 + C_2$ then $\|_C = \|_{C_2} \|_{C_1}$; and that the dependence on C be smooth (this is most easily defined in the context of fiber bundles, which we will cover in a later chapter). If we then choose a frame on $U \subset M$, we have bases for each tangent space that provide isomorphisms $T_pU \cong \mathbb{R}^n$, $T_qU \cong \mathbb{R}^n$. Thus the parallel transporter can be viewed as a map $\|^\lambda{}_\mu \colon \{C\} \to GL(n, \mathbb{R})$ from the set of curves on U to the Lie group $GL(n, \mathbb{R})$; however, it is important to note that the values of $\|^\lambda{}_\mu$ depend upon the choice of frame.

9.1.3 *The covariant derivative*

Recall that the Lie derivative of a vector field $L_v w$ compares the value of w to its value after being "transported" by the local flow of v. Having defined the parallel transporter, we can now consider the **covariant derivative**

$$\nabla_v w \equiv \lim_{\varepsilon \to 0} \frac{1}{\varepsilon} \left(w|_{p+\varepsilon v} - \|_C (w|_p) \right)$$

$$= \lim_{\varepsilon \to 0} \frac{1}{\varepsilon} \left(\|_{-C} (w|_{p+\varepsilon v}) - w|_p \right),$$

where C is an infinitesimal curve starting at p with tangent v. At a point p, $\nabla_v w$ then compares the value of w at $p + \varepsilon v$ to its value at p after being parallel transported to $p + \varepsilon v$, or equivalently in the limit $\varepsilon \to 0$, the value of w at p to its value at $p + \varepsilon v$ after being parallel transported back to p. Recall from Section 6.1.2 that $p + \varepsilon v$ is well-defined in the limit $\varepsilon \to 0$.

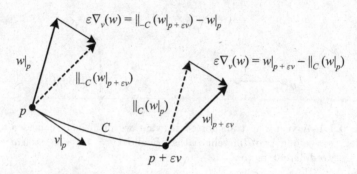

Figure 9.1.2 The covariant derivative $\nabla_v w$ is the difference between a vector field w and its parallel transport in the direction v (for figure conventions, see the box after Figure 6.3.2).

Two properties of $\nabla_v w$ that are easy to verify are that is is linear in v, and that for a function f on M it obeys the rule $\nabla_v (fw) = v(f) w + f \nabla_v (w) = \mathrm{d}f (v) w + f \nabla_v (w)$. As we will see in Section 9.2.1, this is the Leibniz rule for the covariant derivative generalized to the tensor algebra. Note that $\nabla_v w$ is a directional derivative, i.e. it depends only upon the value of v at p; v is in effect used only to choose a direction. In contrast, $L_v w$ requires v to be a vector field, since w is in this case compared to its value after being "transported" by the local flow of v, and so depends on the derivative of v at p.

△ It is important to remember that there is no way to "transport" a vector on a manifold without introducing some extra structure. In particular, recall that the exterior derivative does not compare vectors at all.

Instead of parallel transport, one can consider the covariant derivative as the fundamental structure being added to the manifold. In this case it is useful to define the covariant derivative along a smooth parametrized curve

$C(t)$ by using the tangent to the curve as the direction, i.e.

$$\frac{D}{dt}w \equiv D_t w \equiv \nabla_{\dot{C}(t)} w,$$

where $\dot{C}(t)$ is the tangent to C at t. $D_t w$ is sometimes called the **absolute derivative** (AKA intrinsic derivative) and its definition only requires that w be defined along the curve $C(t)$. We can then define the parallel transport of $w|_p$ along $C(t)$ as the vector field w that satisfies $D_t w = 0$.

$\triangle$ The notation for the absolute derivative is potentially confusing since the implicitly referenced curve $C(t)$ does not appear in the expression $D_t w$.

9.1.4 *The connection*

If we view ∇ as a map from two vector fields v and w to a third vector field $\nabla_v w$, it is called an **affine connection**. Note that since no use has been made of coordinates or frames in the definition of ∇, it is a frame-independent quantity.

Since ∇_v is linear in v, and depends only on its local value, we can regard ∇ as a 1-form on M. If we choose a frame e_μ on M with corresponding dual frame β^μ, we can define the **connection 1-form**

$$\Gamma^\lambda{}_\mu(v) \equiv \beta^\lambda\left(\nabla_v e_\mu\right).$$

$\Gamma^\lambda{}_\mu(v)$ is the λth component of the difference between the frame e_μ and its parallel transport in the direction v.

From its definition, it is clear that $\Gamma^\lambda{}_\mu$ is a frame-dependent object. Under a change of frame $(\gamma^{-1})^\nu{}_\mu$, it is not hard to see that the connection 1-form transforms as

$$\Gamma'^\sigma{}_\tau = \gamma^\sigma{}_\lambda \Gamma^\lambda{}_\mu (\gamma^{-1})^\mu{}_\tau + \gamma^\sigma{}_\lambda d(\gamma^{-1})^\lambda{}_\tau,$$

where the exterior derivative d operates on each of the matrix components $(\gamma^{-1})^\lambda{}_\tau$ as a 0-form. This affine mapping (inhomogeneous transformation) under a change of frame, along with the way that the connection allows us to view an infinitesimal area as an affine space, explains the name "affine connection." It also demonstrates that $\Gamma^\lambda{}_\mu$ cannot be viewed as the components of a tensor, as expected since it is formed from the derivative of the frame.

At a point p, the value of $\Gamma^\lambda{}_\mu(v)$ is an infinitesimal linear transformation on T_pM, so that $\Gamma^\lambda{}_\mu$ is a frame-dependent $gl(n,\mathbb{R})$-valued 1-form. Recalling our notation for algebra-valued forms from Section 3.3.5, we can then write

$$\check{\Gamma}(v)\,\vec{w} \equiv \Gamma^\lambda{}_\mu(v)\,w^\mu e_\lambda = (\nabla_v e_\mu)\,w^\mu,$$

where we view $\vec{w}$ as a $\mathbb{R}^n$-valued 0-form. The vector $\check{\Gamma}(v)\,\vec{w}$ measures the difference between the frame and its parallel transport in the direction v, weighted by the components of w.

$\triangle$ It is important to remember that $\check{\Gamma}(v)\,\vec{w}$ is related to the difference between the frame and its parallel transport, while $\nabla_v w$ measures the difference between w and its parallel transport; thus unlike $\nabla_v w$, $\check{\Gamma}(v)\,\vec{w}$ depends only upon the local value of w, but takes values that are frame-dependent.

$\triangle$ Since we have used the frame to view $\check{\Gamma}$ as a $gl(n,\mathbb{R})$-valued 1-form, i.e. a matrix-valued 1-form, $\vec{w}$ must be viewed as a frame-dependent column vector of components. We could instead view $\check{\Gamma}$ as a $gl(\mathbb{R}^n)$-valued 1-form and $\vec{w}$ as a frame-independent intrinsic vector. In this case the action of $\check{\Gamma}$ on $\vec{w}$ would be frame-independent, but the value of $\check{\Gamma}$ itself would remain frame-dependent. We choose to use matrix-valued forms due to the need below to take the exterior derivative of component functions, but the abstract viewpoint is important to keep in mind when generalizing to fiber bundles in Chapter 10.

Using this notation, we can view a change of frame as a (frame-dependent) $GL(n,\mathbb{R})$-valued 0-form and write the transformation of the connection 1-form under a change of frame as

$$\check{\Gamma}' = \check{\gamma}\check{\Gamma}\check{\gamma}^{-1} + \check{\gamma}\mathrm{d}\check{\gamma}^{-1}.$$

9.1.5 *The covariant derivative in terms of the connection*

$\nabla_v w$ can be written in terms of $\check{\Gamma}$ by using the Leibniz rule from Section 9.1.3 with w^μ as frame-dependent functions:

$$\nabla_v w = \nabla_v \left(w^\mu e_\mu \right)$$
$$= v \left(w^\mu \right) e_\mu + w^\mu \nabla_v \left(e_\mu \right)$$
$$= dw^\mu \left(v \right) e_\mu + \check{\Gamma} \left(v \right) \vec{w}$$
$$\equiv d\vec{w} \left(v \right) + \check{\Gamma} \left(v \right) \vec{w}$$

Here we again view $\vec{w}$ as a $\mathbb{R}^n$-valued 0-form, so that $d\vec{w}(v) \equiv dw^\mu(v) e_\mu$. Thus $d\vec{w}(v)$ is the change in the components of w in the direction v, making it frame-dependent even though w is not. Note that although $\nabla_v w$ is a frame-independent quantity, both terms on the right hand side are frame-dependent. This is depicted in the following figure.

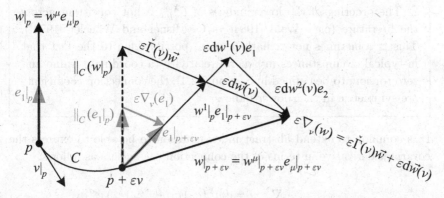

Figure 9.1.3 Relationships between the frame, parallel transport, covariant derivative, and connection for a vector w parallel to e_1 at a point p.

✿ The relation $\nabla_v w = \check{\Gamma}(v)\vec{w} + d\vec{w}(v)$ can be viewed as roughly saying that the change in w under parallel transport is equal to the change in the frame relative to its parallel transport plus the change in the components of w in that frame.

If the 1-form $\Gamma^\lambda{}_\mu(v)$ itself is written using component notation, we arrive at the **connection coefficients**

$$\Gamma^\lambda{}_{\mu\sigma} \equiv \Gamma^\lambda{}_\mu(e_\sigma) = \beta^\lambda \left(\nabla_{e_\sigma} e_\mu \right).$$

$\Gamma^\lambda{}_{\mu\sigma}$ thus measures the λth component of the difference between e_μ and its parallel transport in the direction e_σ.

△ This notation is potentially confusing, as it makes $\Gamma^\lambda{}_{\mu\sigma}$ look like the components of a tensor, which it is not: it is a derivative of the component of the frame indexed by μ, and therefore is not only locally frame-dependent but also depends upon values of the frame at other points, so that it is not a multilinear mapping on its local arguments. Similarly, $d\vec{w}$ looks like a frame-independent exterior derivative, but it is not: it is the exterior derivative of the frame-dependent components of w.

△ The ordering of the lower indices of $\Gamma^\lambda{}_{\mu\sigma}$ is not consistent across the literature (e.g. Wald [1984] vs C. Misner and Wheeler [1973]). This is sometimes not remarked upon, possibly due to the fact that in typical circumstances in general relativity (a coordinate frame and zero torsion, to be defined in Section 9.2.4), the connection coefficients are symmetric in their lower indices.

It is common to extend abstract index notation to be able to express the covariant derivative in terms of the connection coefficients as follows:

$$\nabla_{e_\mu} w = dw^\lambda (e_\mu) e_\lambda + \Gamma^\lambda{}_\sigma (e_\mu) w^\sigma e_\lambda$$
$$\Rightarrow \nabla_a w^b \equiv (\nabla_{e_a} w)^b = e_a (w^b) + \Gamma^b{}_{ca} w^c$$
$$\Rightarrow \nabla_a w^b = \partial_a w^b + \Gamma^b{}_{ca} w^c$$

Here we have also defined $\partial_a f \equiv \partial_{e_a} f = df(e_a) = e_a(f)$. This notation is also sometimes supplemented to use a comma to indicate partial differentiation and a semicolon to indicate covariant differentiation, so that the above becomes

$$w^b{}_{;a} = w^b{}_{,a} + \Gamma^b{}_{ca} w^c.$$

The extension of index notation to derivatives has several potentially confusing aspects:

- ∇_a and ∂_a written alone are not 1-forms
- Greek indices indicate only that a specific basis (frame) has been chosen (Wald [1984] pp. 23-26), but do not distinguish between a general frame, where $\partial_\mu f \equiv df(e_\mu)$, and a coordinate frame, where $\partial_\mu f \equiv \partial f / \partial x^\mu$

- $\nabla_a w^b \equiv (\nabla_{e_a} w)^b$, so since $\nabla_v w$ is linear in v, $\nabla_a w^b$ is in fact a tensor of type $(1,1)$; a more accurate notation might be $(\nabla w)^b{}_a$
- w^b in the expression $\partial_a w^b \equiv dw^b(e_a)$ is not a vector, it is a set of frame-dependent component functions labeled by b whose change in the direction e_a is being measured
- The above means that, consistent with the definition of the connection coefficients, we have $\nabla_a e_b = 0 + e_c \Gamma^c{}_{ba}$, since the components of the frame itself by definition do not change
- As previously noted, neither $\Gamma^b{}_{ca}$ nor $\Gamma^b{}_{ca} w^c$ are tensors

We will nevertheless use this notation for many expressions going forward, as it is frequently used in general relativity.

> △ It is important to remember that expressions involving ∇_a, ∂_a, and $\Gamma^c{}_{ba}$ must be handled carefully, as none of these are consistent with the original concept of indices denoting tensor components.

> △ Some texts will distinguish between the labels of basis vectors and abstract index notation by using expressions such as $(e_i)^a$. We will not follow this practice, as it makes difficult the convenient method of matching indexes in expressions such as $\partial_a w^b \equiv dw^b(e_a)$.

> △ If we choose coordinates x^μ and use a coordinate frame so that $\partial_\mu \equiv \partial/\partial x^\mu$, we have the usual relation $\partial_\mu \partial_\nu f = \partial_\nu \partial_\mu f$. However, this is not necessarily implied by the Greek indices alone, which only indicate that a particular frame has been chosen. For index notation in general, mixed partials do not commute, since $\partial_a \partial_b f - \partial_b \partial_a f = e_a(e_b(f)) - e_b(e_a(f)) = [e_a, e_b](f) = [e_a, e_b]^c \partial_c f$, which only vanishes in a holonomic frame.

9.1.6 *The parallel transporter in terms of the connection*

We can also consider the parallel transport of a vector w along an infinitesimal curve C with tangent v. Referring to Figure 9.1.3, we see that to order ε the components w^μ transform according to

$$\|^\lambda{}_\mu (C) w^\mu = w^\lambda - \varepsilon \Gamma^\lambda{}_\mu (v) w^\mu,$$

where v is tangent to the curve C, and these components are with respect to the frame at the new point after infinitesimal parallel transport. Using this relation, we can build up a frame-dependent expression for the parallel transporter for finite C by multiplying terms $(1 - \varepsilon\Gamma|_p)$ where $\Gamma|_p$ is used to denote the matrix $\Gamma^\lambda{}_\mu (v|_p)$ evaluated on the tangent $v|_p$ at successive points p along C. The limit of this process is the **path-ordered exponential**

$$\|^\lambda{}_\mu (C) = \lim_{\varepsilon \to 0} (1 - \varepsilon\Gamma|_{q-\varepsilon}) (1 - \varepsilon\Gamma|_{q-2\varepsilon}) \cdots (1 - \varepsilon\Gamma|_{p+\varepsilon}) (1 - \varepsilon\Gamma|_p)$$

$$\equiv P\exp\left(-\int_C \Gamma^\lambda{}_\mu\right),$$

whose definition is based on the expression for the exponential

$$e^x = \lim_{n \to \infty} \left(1 + \frac{x}{n}\right)^n = \lim_{\varepsilon \to \infty} (1 + \varepsilon x)^{1/\varepsilon}.$$

Note that the above expression for $\|^\lambda{}_\mu (C)$ exponentiates frame-dependent values in $gl(n, \mathbb{R})$ to yield a frame-dependent value in $GL(n, \mathbb{R})$.

9.1.7 *Geodesics and normal coordinates*

Following the example of the Lie derivative (Section 6.3.2), we can consider parallel transport of a vector v in the direction v as generating a local flow. More precisely, for any vector v at a point $p \in M$, there is a curve $\phi_v(t)$, unique for some $-\varepsilon < t < \varepsilon$, such that $\phi_v(0) = p$ and $\dot{\phi}_v(t) = \|_\phi (v)$, the last expression indicating that the tangent to ϕ_v at t is equal to the parallel transport of v along ϕ_v from $\phi_v(0)$ to $\phi_v(t)$. This curve is called a **geodesic**, and its tangent vectors are all parallel transports of each other. This means that for all tangent vectors v to the curve, $\nabla_v v = 0$, so that geodesics are "the closest thing to straight lines" on a manifold with parallel transport.

Now following the example of Lie groups (Section 7.2.2), we can define the exponential map at p to be $\exp(v) \equiv \phi_v(1)$, which will be well-defined for values of v around the origin that map to some $U \subset M$ containing p. Finally, choosing a basis for T_pU provides an isomorphism $T_pU \cong \mathbb{R}^n$, allowing us to define **geodesic normal coordinates** (AKA normal coordinates) $\exp^{-1} \colon U \to \mathbb{R}^n$. It can be shown (see Kobayashi and Nomizu [1963] Vol. 1 pp148-149) that in a coordinate frame at the origin p of geodesic normal coordinates, we have $\Gamma^\lambda{}_{\mu\sigma} = -\Gamma^\lambda{}_{\sigma\mu}$; this implies that for zero torsion (to be defined in Section 9.2.4), the connection coefficients vanish at p.

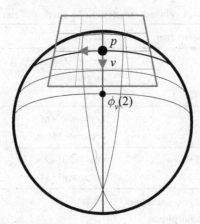

Figure 9.1.4 Geodesic normal coordinates at p map points on a manifold to vectors at p tangent to the geodesic passing through both points. In the figure $\exp(2v) = \phi_v(2)$, so the coordinate of the point $\phi_v(2) \in M$ is $2v \in T_pM$.

This is the third time we have utilized the concept of the flows (AKA integral curves, field lines, streamlines, trajectories, orbits) of vector fields. The relationships between these three situations are summarized below.

Table 9.1.2 Flows and related constructs in three different contexts.

	Lie derivative	Lie group	Covariant derivative
Added structure on M	Vector field v	Group structure on points	Parallel transport $\parallel_C (v)$
Vector field on M	v	Left-invariant vector fields A	Tangents to geodesics
Flow of vector field	Local flow $v_p(t)$	One-parameter subgroup $\phi_A(t)$	Geodesics $\phi_v(t)$
Exponential map of flow	Local one-parameter diffeomorphism $\Phi_t(v) \equiv v_p(t)$	$e^A \equiv \phi_A(1)$	$\exp(v) \equiv \phi_v(1)$
Diffeo-morphism of exp	Local in general, global if v is complete	Local to identity	Local to origin of exp
Vector transport by flow	Tangent map (only along v)	Tangent map along A	Parallel translation along geodesics $\parallel_\phi (v)$
Vector derivative from transport	$L_v w$ is the difference between w and its transport by the local flow of v	$[A, B]$ is the difference between B and its transport by the local flow of A	$\nabla_v w$ is the difference between w and its parallel transport in the direction v

9.1.8 *Summary*

In general, a "manifold with connection" is one with an additional structure that "connects" the different tangent spaces of the manifold to one another in a linear fashion. Specifying any one of the above connection quantities, the covariant derivative, or the parallel transporter equivalently determines this structure. The following tables summarize the situation.

Table 9.1.3 Constructions related to the connection.

Construct	Argument(s)	Value	Dependencies
$\parallel_C$	$v \in T_p M$	$\parallel_C (v) \in T_q M$	Path C from p to q
$\parallel^\lambda{}_\mu$	Path C	$\parallel^\lambda{}_\mu (C) \in GL$	Frame on M
∇_v	$w \in TM$	$\nabla_v w \in T_p M$	$v \in T_p M$
∇	$v \in T_p M,\ w \in TM$	$\nabla_v w \in T_p M$	None
$\Gamma^\lambda{}_\mu$	$v \in T_p M$	$\Gamma^\lambda{}_\mu (v) \in gl$	Frame on M
$\check{\Gamma} (v)$	$\vec{w} \in T_p M$	$\check{\Gamma} (v)\, \vec{w} \in T_p M$	Frame on M, $v \in T_p M$
$\Gamma^\lambda{}_{\mu\sigma}$	None	Connection coefficient	Frame on M

Notes: Each construct above is considered at a point p; to determine a manifold with connection it must be defined for every point in M.

Below we review the intuitive meanings of the various vector derivatives.

Table 9.1.4 Definitions and meanings of vector derivatives.

Vector derivative	Meaning
$L_v w \equiv \lim_{\varepsilon \to 0} \left(w\vert_{p+\varepsilon v} - \mathrm{d}\Phi_\varepsilon \left(w\vert_p \right) \right)/\varepsilon$	The difference between w and its transport by the local flow of v.
$\nabla_v w \equiv \lim_{\varepsilon \to 0} \left(w\vert_{p+\varepsilon v} - \Vert_C \left(w\vert_p \right) \right)/\varepsilon$	The difference between w and its parallel transport in the direction v.
$\frac{\mathrm{D}}{\mathrm{d}t} w \equiv \mathrm{D}_t w \equiv \nabla_{\dot{C}(t)} w$	The difference between w and its parallel transport in the direction tangent to $C(t)$.
$\Gamma^\lambda{}_\mu (v) \equiv \beta^\lambda \left(\nabla_v e_\mu \right)$	The λth component of the difference between e_μ and its parallel transport in the direction v.
$\check{\Gamma}(v) \equiv \nabla_v (T_p M)$	The infinitesimal linear transformation on the tangent space that takes the parallel transported frame to the frame in the direction v.
$\check{\Gamma}(v)\,\vec{w} \equiv \Gamma^\lambda{}_\mu (v)\, w^\mu e_\lambda = \left(\nabla_v e_\mu \right) w^\mu$	The difference between the frame and its parallel transport in the direction v, weighted by the components of w.
$\Gamma^\lambda{}_{\mu\sigma} \equiv \Gamma^\lambda{}_\mu (e_\sigma) = \beta^\lambda \left(\nabla_\sigma e_\mu \right)$	The λth component of the difference between e_μ and its parallel transport in the direction e_σ.
$\mathrm{d}\vec{w}(v) \equiv \mathrm{d}w^\mu (v)\, e_\mu$	The change in the frame-dependent components of w in the direction v.
$\partial_a w^b \equiv \mathrm{d}w^b (e_a)$	The change in the bth frame-dependent component of w in the direction e_a.
$\nabla_a w^b \equiv \left(\nabla_{e_a} w \right)^b$	The bth component of the difference between w and its parallel transport in the direction e_a.

Other quantities in terms of the connection:

- $\nabla_v w = \mathrm{d}\vec{w}(v) + \check{\Gamma}(v)\,\vec{w}$
- $\nabla_a w^b = \partial_a w^b + \Gamma^b{}_{ca} w^c$
- $\Vert^\lambda{}_\mu (C)\, w^\mu = w^\lambda - \varepsilon \Gamma^\lambda{}_\mu (v)\, w^\mu$ (for infinitesimal C with tangent v)
- $\Vert^\lambda{}_\mu (C)\, w^\mu = P\exp\left(-\int_C \Gamma^\lambda{}_\mu \right) w^\mu$

9.2 Manifolds with connection

All of the above constructs used to define a manifold with connection manipulate vectors, which means they can be naturally extended to operate on arbitrary tensor fields on M. This is the usual approach taken in general relativity; however, one can alternatively focus on k-forms on M, an approach that generalizes more directly to gauge theories in physics. This viewpoint is sometimes called the **Cartan formalism**. We will cover both approaches.

Since many of the standard texts in this area only cover one of these viewpoints, and in addition often assume a coordinate frame, a metric, and/or zero torsion (to be defined in Section 9.2.4), we include a bit more calculational detail here than in other sections.

> △ Note that a manifold with connection includes no concept of length or distance (a metric). It is important to remember that unless noted, nothing in this section depends upon this extra structure.

9.2.1 *The covariant derivative on the tensor algebra*

If we define the covariant derivative of a function to coincide with the normal derivative, i.e. $\nabla_a f \equiv \partial_a f$, then we can use the Leibniz rule to define the covariant derivative of a 1-form. This is sometimes described as making the covariant derivative "commute with contractions," where for a 1-form φ and a vector v we require

$$\nabla_a \left(\varphi_b v^b \right) \equiv \left(\nabla_a \varphi_b \right) v^b + \varphi_b \left(\nabla_a v^b \right)$$
$$= \left(\nabla_a \varphi_b \right) v^b + \varphi_b \left(\partial_a v^b + \Gamma^b{}_{ca} v^c \right).$$

At the same time, choosing a frame and treating φ_b and v^b as frame-dependent functions on M, we have

$$\nabla_a \left(\varphi_b v^b \right) \equiv \partial_a \left(\varphi_b v^b \right)$$
$$= \left(\partial_a \varphi_b \right) v^b + \varphi_b \left(\partial_a v^b \right),$$

so that equating the two we arrive at

$$\nabla_a \varphi_b \equiv \partial_a \varphi_b - \Gamma^c{}_{ba} \varphi_c.$$

As with vectors, the partial derivative $\partial_a \varphi_b$ acts upon the frame-dependent components of the 1-form.

We can then extend the covariant derivative to be a derivation on the tensor algebra by following the above logic for each covariant and contravariant component:

$$\nabla_a T^{b_1 \ldots b_m}{}_{c_1 \ldots c_n} \equiv \partial_a T^{b_1 \ldots b_m}{}_{c_1 \ldots c_n}$$

$$+ \sum_{j=1}^{m} \Gamma^{b_j}{}_{da} T^{b_1 \ldots b_{j-1} d b_{j+1} \ldots b_m}{}_{c_1 \ldots c_n}$$

$$- \sum_{j=1}^{n} \Gamma^{d}{}_{c_j a} T^{b_1 \ldots b_m}{}_{c_1 \ldots c_{j-1} d c_{j+1} \ldots c_n}$$

Note that since the covariant derivative of a 0-form is $\nabla_a f = \partial_a f = \partial_{e_a} f = e_a(f)$, we then have $\nabla_v f = v^a \nabla_a f = v^a e_a(f) = v(f)$.

The concept of parallel transport along a curve C can be extended to the tensor algebra as well, by parallel transporting all vector arguments backwards to the starting point of C, applying the tensor, then parallel transporting the resulting vectors forward to the endpoint of C. So for example the parallel transport of a tensor $T^a{}_b$ is defined as

$$\|_C (T^a{}_b) \equiv \|^a{}_c (C) \, T^c{}_d \, \|^d{}_b (-C)$$
$$= (1 - \varepsilon \Gamma^a{}_c (v)) \, T^c{}_d \left(1 + \varepsilon \Gamma^d{}_b (v)\right),$$

where for infinitesimal C with tangent v we have $\|_C^{-1} = \|_{-C} = 1 + \varepsilon \check{\Gamma}(v)$ since $\|_C = 1 - \varepsilon \check{\Gamma}(v)$.

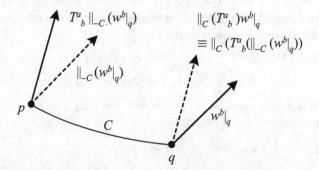

Figure 9.2.1 The parallel transport of a tensor can be defined by parallel transporting all vector arguments backwards to the starting point, applying the tensor, then parallel transporting the resulting vectors forward to the endpoint.

With this definition, the covariant derivative $\nabla_a T$ can be viewed as "the difference between T and its parallel transport in the direction e_a."

△ It can sometimes be confusing when using the extended covariant derivative as to what type of tensor it is being applied to. For example, w^b in the expression $\partial_a w^b$ is not a vector, it is a set of frame-dependent functions labeled by b; yet this expression can in theory also be written $\nabla_a w^b$, in which case there is no indication that the covariant derivative is acting on these functions instead of the vector w^b.

△ When the covariant derivative is used as a derivation on the tensor algebra, care must be taken with relations, since their forms can change considerably based upon what arguments are applied and whether index notation is used. In particular, $(\nabla_a \nabla_b - \nabla_b \nabla_a)f = \nabla_a(\partial_b f) - \nabla_b(\partial_a f)$ is not a "mixed partials" expression, since $(\partial_a f)$ is a 1-form. And as we will see, $(\nabla_a \nabla_b - \nabla_b \nabla_a)f$ is a different construction than $(\nabla_a \nabla_b - \nabla_b \nabla_a)w^c$, which is different from $(\nabla_u \nabla_v - \nabla_v \nabla_u)w$. It is important to realize that an expression such as $\nabla_a \nabla_b - \nabla_b \nabla_a$ without context has no unambiguous meaning.

△ It is important to remember that since expressions like $\partial_a w^b$ and $\Gamma^c{}_{ba}$ are not tensors, applying ∇_d to them is not well-defined (unless we consider them as arrays of functions and are applying $\nabla_d = \partial_d$).

9.2.2 The exterior covariant derivative of vector-valued forms

A vector field w on M can be viewed as a vector-valued 0-form. As noted previously, the covariant derivative $\nabla_v w$ is linear in v and depends only on its local value, and so can be viewed as a vector-valued 1-form $D\vec{w}(v) \equiv \nabla_v w$. $D\vec{w}$ is called the **exterior covariant derivative** of the vector-valued 0-form $\vec{w}$. This definition is then extended to vector-valued k-forms $\vec{\varphi}$ by following the example of the exterior derivative d in Section 6.3.5:

$$\mathrm{D}\vec{\varphi}\,(v_0,\dots,v_k)$$

$$\equiv \sum_{j=0}^{k}(-1)^j\,\nabla_{v_j}\left(\vec{\varphi}\,(v_0,\dots,v_{j-1},v_{j+1},\dots,v_k)\right)$$

$$+\sum_{i<j}(-1)^{i+j}\,\vec{\varphi}\left([v_i,v_j]\,,v_0,\dots,v_{i-1},v_{i+1},\dots,v_{j-1},v_{j+1},\dots,v_k\right)$$

For example, if $\vec{\varphi}$ is a vector-valued 1-form, we have

$$\mathrm{D}\vec{\varphi}\,(v,w)=\nabla_v\vec{\varphi}\,(w)-\nabla_w\vec{\varphi}\,(v)-\vec{\varphi}\,([v,w])\,.$$

So while the first term of $\mathrm{d}\varphi$ takes the difference between the scalar values of $\varphi(w)$ along v, the first term of $\mathrm{D}\vec{\varphi}$ takes the difference between the vector values of $\vec{\varphi}(w)$ along v after parallel transporting them to the same point (which is required to compare them). At a point p, $\mathrm{D}\vec{\varphi}(v,w)$ can thus be viewed as the "sum of $\vec{\varphi}$ on the boundary of the surface defined by its arguments after being parallel transported back to p," and if we use $\|_{\varepsilon v}$ to denote parallel transport along an infinitesimal curve with tangent v, we can write

$$\varepsilon^2\mathrm{D}\vec{\varphi}\,(v,w)=\|_{-\varepsilon v}\vec{\varphi}\,(\varepsilon w\,|_{p+\varepsilon v})-\vec{\varphi}\,(\varepsilon w\,|_p)$$
$$-\|_{-\varepsilon w}\vec{\varphi}\,(\varepsilon v\,|_{p+\varepsilon w})+\vec{\varphi}\,(\varepsilon v\,|_p)$$
$$-\vec{\varphi}\left(\varepsilon^2\,[v,w]\right)\,.$$

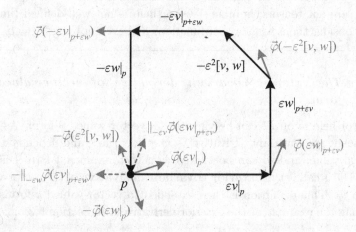

Figure 9.2.2 The exterior covariant derivative $\mathrm{D}\vec{\varphi}(v,w)$ sums the vectors $\vec{\varphi}$ along the boundary of the surface defined by v and w by parallel transporting them to the same point. Note that the "completion of the parallelogram" $[v,w]$ is already of order ε^2, so its parallel transport has no effect to this order.

From its definition, it is clear that $D\vec{\varphi}$ is a frame-independent quantity. In terms of the connection, we must consider $\vec{w}$ as a frame-dependent $\mathbb{R}^n$-valued 0-form, so that

$$D\vec{w}(v) = \nabla_v w = d\vec{w}(v) + \check{\Gamma}(v)\,\vec{w}.$$

For a $\mathbb{R}^n$-valued k-form $\vec{\varphi}$ we find that

$$D\vec{\varphi} = d\vec{\varphi} + \check{\Gamma} \wedge \vec{\varphi},$$

where the exterior derivative is defined to apply to the frame-dependent components, i.e. $d\vec{\varphi}(v_0 \ldots v_k) \equiv d\varphi^\mu(v_0 \ldots v_k)e_\mu$. Recall that $\check{\Gamma}$ is a $gl(n,\mathbb{R})$-valued 1-form, so that for example if $\vec{\varphi}$ is a $\mathbb{R}^n$-valued 1-form then $(\check{\Gamma} \wedge \vec{\varphi})(v,w) \equiv \check{\Gamma}(v)\,\vec{\varphi}(w) - \check{\Gamma}(w)\,\vec{\varphi}(v) = \Gamma^\lambda{}_\mu(v)\,\varphi^\mu(w) - \Gamma^\lambda{}_\mu(w)\,\varphi^\mu(v)$.

$\triangle$ As with the covariant derivative, it is important to remember that $D\vec{\varphi}$ is frame-independent while $d\vec{\varphi}$ and $\check{\Gamma}$ are not.

The set of vector-valued forms can be viewed as an infinite-dimensional algebra by defining multiplication via the vector field commutator; it turns out that D does not satisfy the Leibniz rule in this algebra and so is not a derivation. However, following the above reasoning one can extend the definition of D to the algebra of tensor-valued forms, or the subset of anti-symmetric tensor-valued forms; D then is a derivation with respect to the tensor product in the former case and a graded derivation with respect to the exterior product in the latter case. We will not pursue either of these two generalizations.

9.2.3 The exterior covariant derivative of algebra-valued forms

Recalling from Section 9.2.1 the definition of parallel transport of a tensor, we can view a $gl(n,\mathbb{R})$-valued 0-form $\check{\Theta}$ as a tensor field of type $(1,1)$, so that the infinitesimal parallel transport of $\check{\Theta}$ along C with tangent v is

$$\|_C(\check{\Theta}) = \left(1 - \varepsilon\check{\Gamma}(v)\right)\check{\Theta}\left(1 + \varepsilon\check{\Gamma}(v)\right).$$

We can now follow the reasoning used to define the covariant derivative of a vector in terms of the connection

$$\nabla_v w \equiv \lim_{\varepsilon \to 0} \frac{1}{\varepsilon} \left(w \mid_{p+\varepsilon v} - \parallel_C (w \mid_p) \right)$$

$$= \lim_{\varepsilon \to 0} \frac{1}{\varepsilon} \left(\vec{w} \mid_{p+\varepsilon v} - \left(1 - \varepsilon \check{\Gamma}(v) \right) \vec{w} \mid_p \right)$$

$$= \lim_{\varepsilon \to 0} \frac{1}{\varepsilon} \left(w^\mu \mid_{p+\varepsilon v} - w^\mu \mid_p + \varepsilon \Gamma^\mu{}_\lambda(v) w^\lambda \mid_p \right) e_\mu \mid_{p+\varepsilon v}$$

$$= d\vec{w}(v) + \check{\Gamma}(v) \vec{w}$$

to give the covariant derivative of a $gl(n, \mathbb{R})$-valued 0-form

$$\nabla_v \check{\Theta} \equiv \lim_{\varepsilon \to 0} \frac{1}{\varepsilon} \left(\check{\Theta} \mid_{p+\varepsilon v} - \parallel_C (\check{\Theta} \mid_p) \right)$$

$$= \lim_{\varepsilon \to 0} \frac{1}{\varepsilon} \left(\check{\Theta} \mid_{p+\varepsilon v} - \left(1 - \varepsilon \check{\Gamma}(v) \right) \check{\Theta} \mid_p \left(1 + \varepsilon \check{\Gamma}(v) \right) \right)$$

$$= d\check{\Theta}(v) + \check{\Gamma}(v) \check{\Theta} - \check{\Theta} \check{\Gamma}(v)$$

$$= d\check{\Theta}(v) + \left[\check{\Gamma}, \check{\Theta} \right](v)$$

$$= d\check{\Theta}(v) + \left(\check{\Gamma}[\wedge]\check{\Theta} \right)(v).$$

Here we have only kept terms to order ε, followed previous convention to define $d\check{\Theta}(v) \equiv d\Theta^\mu{}_\lambda \beta^\lambda e_\mu$, and defined the Lie commutator $[\check{\Gamma}, \check{\Theta}]$ in terms of the multiplication of the $gl(n, \mathbb{R})$-valued forms $\check{\Gamma}$ and $\check{\Theta}$, which recalling our notation from Section 3.3.5 as a 1-form is equivalent to $\check{\Gamma}[\wedge]\check{\Theta}$. $\nabla_v \check{\Theta}$ is then "the difference between the linear transformation $\check{\Theta}$ and its parallel transport in the direction v."

The above definition of the covariant derivative can then be extended to arbitrary $gl(n, \mathbb{R})$-valued k-forms by defining

$$D\check{\Theta} \equiv d\check{\Theta} + \check{\Gamma}[\wedge]\check{\Theta},$$

which can be shown to be equivalent to the construction used for $\mathbb{R}^n$-valued k-forms in Section 9.2.2. For example, for a $gl(n, \mathbb{R})$-valued 1-form $\check{\Theta}$, we have $D\check{\Theta}(v, w) \equiv \nabla_v \check{\Theta}(w) - \nabla_w \check{\Theta}(v) - \check{\Theta}([v, w])$, with the covariant derivatives acting on the value of $\check{\Theta}$ as a tensor of type $(1, 1)$. So at a point p, $D\check{\Theta}(v, w)$ can be viewed as the "sum of $\check{\Theta}$ on the boundary of the surface defined by its arguments after being parallel transported back to p." With respect to the set of $gl(n, \mathbb{R})$-valued forms under the exterior product using the Lie commutator $[\wedge]$, D is a graded derivation and for a $gl(n, \mathbb{R})$-valued k-form $\check{\Theta}$ satisfies the Leibniz rule $D(\check{\Theta}[\wedge]\check{\Psi}) = D\check{\Theta}[\wedge]\check{\Psi} + (-1)^k \check{\Theta}[\wedge]D\check{\Psi}$.

9.2.4 *Torsion*

Given a frame e_μ, we can view the dual frame β^μ as a vector-valued 1-form that simply returns its vector argument: $\vec{\beta}(v) \equiv \beta^\mu(v) e_\mu = v$. Clearly this is a frame-independent object. The **torsion** is then defined to be the exterior covariant derivative

$$\vec{T} \equiv D\vec{\beta}.$$

In terms of the connection, we must consider $\vec{\beta}$ as a frame-dependent $\mathbb{R}^n$-valued 1-form, which gives us the torsion as a $\mathbb{R}^n$-valued 2-form

$$\vec{T} = d\vec{\beta} + \check{\Gamma} \wedge \vec{\beta}.$$

This definition of $\vec{T}$ is sometimes called **Cartan's first structure equation**.

In terms of the covariant derivative, the torsion 2-form is

$$\vec{T}(v,w) \equiv \nabla_v \left(\vec{\beta}(w) \right) - \nabla_w \left(\vec{\beta}(v) \right) - \vec{\beta}([v,w])$$
$$= \nabla_v w - \nabla_w v - [v,w].$$

For a torsion-free connection in a holonomic frame, we then have $\nabla_\sigma e_\mu = \nabla_\mu e_\sigma$, which means that the connection coefficients are symmetric in their lower indices, i.e. $\Gamma^\lambda{}_{\mu\sigma} \equiv \beta^\lambda (\nabla_\sigma e_\mu) = \beta^\lambda (\nabla_\mu e_\sigma) = \Gamma^\lambda{}_{\sigma\mu}$. For this reason, a torsion-free connection is also called a **symmetric connection**.

From the definition in terms of the exterior covariant derivative, we can view the torsion as the "sum of the boundary vectors of the surface defined by its arguments after being parallel transported back to p," i.e. the torsion measures the amount by which the boundary of a loop fails to close after being parallel transported. From the definition in terms of the covariant derivative, we arrive in the figure below at another interpretation where, like the Lie derivative $L_v w$, $\vec{T}(v,w)$ "completes the parallelogram" formed by its vector arguments, but this parallelogram is formed by parallel transport instead of local flow. Note however that the torsion vector has the opposite sign as the Lie derivative.

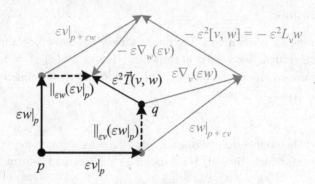

Figure 9.2.3 The torsion vector $\vec{T}(v, w)$, constructed above starting at the point q, "completes the parallelogram" formed by parallel transport. $\|_{\varepsilon v}$ denotes parallel transport along an infinitesimal curve with tangent v.

Zero torsion then means that moving infinitesimally along v followed by the parallel transport of w is the same as moving infinitesimally along w followed by the parallel transport of v. Non-zero torsion signifies that "a loop made of parallel transported vectors is not closed."

As this geometric interpretation suggests, and as is evident from the expression $\vec{T} \equiv \mathrm{D}\vec{\beta}$, one can verify algebraically that despite being defined in terms of derivatives $\vec{T}(v, w)$ in fact only depends on the local values of v and w, and thus can be viewed as a tensor of type $(1, 2)$:

$$T^c{}_{ab} v^a w^b \equiv v^a \nabla_a w^c - w^a \nabla_a v^c - [v, w]^c$$

Another relation can be obtained for the torsion tensor by applying its vector value to a function f before moving into index notation:

$$\vec{T}(v, w)(f) \equiv (\nabla_v w)(f) - (\nabla_w v)(f) - [v, w](f)$$
$$\Rightarrow T^c{}_{ab} v^a w^b \nabla_c f = \left(v^a \nabla_a w^b\right) \nabla_b f - \left(w^b \nabla_b v^a\right) \nabla_a f$$
$$- \left[v^a \nabla_a \left(w^b \nabla_b f\right) - w^b \nabla_b \left(v^a \nabla_a f\right)\right]$$
$$\Rightarrow T^c{}_{ab} \nabla_c f = \nabla_b \nabla_a f - \nabla_a \nabla_b f$$

Here we have used the Leibniz rule and recalled that $v(f) = \nabla_v f = v^a \nabla_a f$ and $[v, w](f) = v(w(f)) - w(v(f))$. In terms of the connection coefficients $\Gamma^c{}_{ab} = \beta^c \nabla_b e_a$ we have

$$T^c{}_{ab} = \beta^c \vec{T}(e_a, e_b)$$
$$= \beta^c \nabla_a e_b - \beta^c \nabla_b e_a - \beta^c [e_a, e_b]$$
$$= \Gamma^c{}_{ba} - \Gamma^c{}_{ab} - [e_a, e_b]^c.$$

△ Note that zero torsion thus always means that $\nabla_a \nabla_b f = \nabla_b \nabla_a f$ (and $[v, w] = L_v w = \nabla_v w - \nabla_w v$), but it only means $\Gamma^\lambda{}_{\mu\sigma} = \Gamma^\lambda{}_{\sigma\mu}$ in a holonomic frame.

In the previous figure, the failure of the parallel transported vectors to meet can be viewed as either due to their lengths changing or due to their being rotated out of the plane of the figure. As we will see, the latter interpretation is more relevant for Riemannian manifolds, where parallel transport leaves lengths invariant. In Einstein-Cartan theory in physics, non-zero torsion is associated with spin in matter. An example along these lines that highlights the rotation aspect of torsion is Euclidean $\mathbb{R}^3$ with parallel transport defined by translation, except in the x direction where parallel transport rotates a vector clockwise by an angle proportional to the distance transported. As we will see in the next section, this parallel transport has torsion but no curvature.

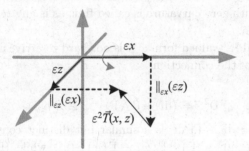

Figure 9.2.4 An example of non-zero torsion suggestive of spin.

Zero torsion means that $L_v w = [v, w] = \nabla_v w - \nabla_w v$ due to the symmetric connection coefficients canceling. This extends to the Lie derivative of a general tensor, so that in the case of zero torsion we have

$$
\begin{aligned}
L_v T^{a_1 \ldots a_m}{}_{b_1 \ldots b_n} =\ & v^c \nabla_c T^{a_1 \ldots a_m}{}_{b_1 \ldots b_n} \\
& - \sum_{j=1}^{m} \left(\nabla_c v^{a_j} \right) T^{a_1 \ldots a_{j-1} c a_{j+1} \ldots a_m}{}_{b_1 \ldots b_n} \\
& + \sum_{j=1}^{n} \left(\nabla_{b_j} v^c \right) T^{a_1 \ldots a_m}{}_{b_1 \ldots b_{j-1} c b_{j+1} \ldots b_n}.
\end{aligned}
$$

9.2.5   *Curvature*

The exterior covariant derivative D parallel transports its values on the boundary before summing them, and therefore we do not expect it to mimic the property $\mathrm{d}^2 = 0$. Indeed it does not; instead, for a vector field w viewed as a vector-valued 0-form $\vec{w}$, we have

$$\left(\mathrm{D}^2\vec{w}\right)(u,v) \equiv \check{R}(u,v)\,\vec{w} = \nabla_u\nabla_v w - \nabla_v\nabla_u w - \nabla_{[u,v]}w,$$

which defines the **curvature 2-form** $\check{R}$, which is $gl(\mathbb{R}^n)$-valued. From its definition, $\check{R}\vec{w}$ is a frame-independent quantity, and thus if $\vec{w}$ is considered as a vector-valued 0-form, $\check{R}$ is frame-independent as well. In the (more common) case that we view $\vec{w}$ as a frame-dependent $\mathbb{R}^n$-valued 0-form, $\check{R}$ must be considered to be $gl(n,\mathbb{R})$-valued, and is thus a frame-dependent matrix, transforming under a (frame-dependent) $GL(n,\mathbb{R})$-valued 0-form $\check{\gamma}^{-1}$ change of frame like

$$\check{R}' = \check{\gamma}\check{R}\check{\gamma}^{-1}.$$

A connection with zero curvature is called **flat**, as is any region of M with a flat connection.

For a general $\mathbb{R}^n$-valued form $\vec{\varphi}$ it is not hard to arrive at an expression for $\check{R}$ in terms of the connection:

$$\mathrm{D}^2\vec{\varphi} = \left(\mathrm{d}\check{\Gamma} + \check{\Gamma}\wedge\check{\Gamma}\right)\wedge\vec{\varphi} \equiv \check{R}\wedge\vec{\varphi}$$

Note that $\mathrm{D}\check{\Gamma} = \mathrm{d}\check{\Gamma} + \check{\Gamma}[\wedge]\check{\Gamma}$ is a similar but distinct construction, since e.g. $(\check{\Gamma}\wedge\check{\Gamma})(v,w) = \check{\Gamma}(v)\check{\Gamma}(w) - \check{\Gamma}(w)\check{\Gamma}(v)$, while $(\check{\Gamma}[\wedge]\check{\Gamma})(v,w) = [\check{\Gamma}(v),\check{\Gamma}(w)] - [\check{\Gamma}(w),\check{\Gamma}(v)] = 2(\check{\Gamma}\wedge\check{\Gamma})(v,w)$. Thus we have

$$\check{R} \equiv \mathrm{d}\check{\Gamma} + \check{\Gamma}\wedge\check{\Gamma}$$
$$= \mathrm{d}\check{\Gamma} + \frac{1}{2}\check{\Gamma}[\wedge]\check{\Gamma}.$$

The definition of $\check{R}$ in terms of $\check{\Gamma}$ is sometimes called **Cartan's second structure equation**. An immediate property from the definition of $\check{R}$ is $\check{R}(u,v) = -\check{R}(v,u)$, which allows us to write e.g. for a vector-valued 1-form $\vec{\varphi}$

$$\left(\mathrm{D}^2\vec{\varphi}\right)(u,v,w) \equiv \left(\check{R}\wedge\vec{\varphi}\right)(u,v,w)$$
$$= \check{R}(u,v)\,\vec{\varphi}(w) + \check{R}(v,w)\,\vec{\varphi}(u) + \check{R}(w,u)\,\vec{\varphi}(v).$$

Constructing the same picture as we did for the double exterior derivative in Section 6.3.4, we put $\mathrm{D}^2\vec{w} \equiv \mathrm{D}\vec{\varphi}$, where $\vec{\varphi}(v) \equiv \mathrm{D}\vec{w}(v) = \nabla_v w$.

Expanding both derivatives in terms of parallel transport, we find in the following figure that as we sum values around the boundary of the surface defined by its arguments, D^2 fails to cancel the endpoint and starting point at the far corner. Examining the values of these non-canceling points, we can view the curvature as "the difference between w when parallel transported around the two opposite edges of the boundary of the surface defined by its arguments."

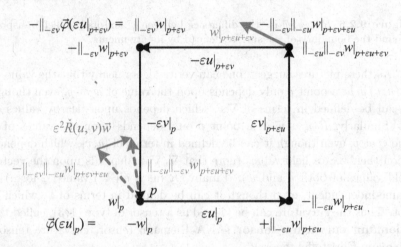

Figure 9.2.5 $\check{R}(u, v)\, \vec{w} = (D^2 \vec{w})(u, v)$ is "the difference between w when parallel transported around the two opposite edges of the boundary of the surface defined by its arguments." In the figure we assume vanishing Lie bracket for simplicity, so that $v|_{p+\varepsilon u+\varepsilon v} = v|_{p+\varepsilon v+\varepsilon u}$.

In terms of the connection, we can use the path integral formulation to examine the parallel transporter around the closed path $L \equiv \partial S$ defined by the surface $S \equiv (\varepsilon u \wedge \varepsilon v)$ to order ε^2. This calculation after some work (see Göckeler and Schücker [1987] pp. 51-53) yields

$$\|_L (w) = P\exp\left(-\int_L \check{\Gamma}\right) \vec{w} = w - \int_S \left(d\check{\Gamma} + \check{\Gamma} \wedge \check{\Gamma}\right) \vec{w} = w - \varepsilon^2 \check{R}(u, v)\, \vec{w},$$

where we have dropped the indices since L is a closed path and thus $\|_L$ is basis-independent. Thus the curvature can be viewed as "the difference between w and its parallel transport around the boundary of the surface defined by its arguments."

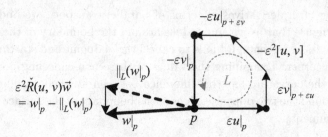

Figure 9.2.6 $\check{R}(u,v)\vec{w}$ is "the difference between w and its parallel transport around the boundary of the surface defined by its arguments."

As these pictures suggest, one can verify algebraically that the value of $\check{R}(u,v)\vec{w}$ at a point p only depends upon the value of w at p, even though it can be defined in terms of ∇w, which depends upon nearby values of w. Similarly, $\check{R}(u,v)\vec{w}$ at a point p only depends upon the values of u and v at p, even though it can be defined in terms of $[u,v]$, which depends upon their vector field values (note that $\nabla_u \nabla_v w$ depends upon the vector field values of both v and w). Finally, $\check{R}$ (as a $gl(\mathbb{R}^n)$-valued 2-form) is frame-independent, even though it can be defined in terms of $\check{\Gamma}$, which is not. Thus the curvature can be viewed as a tensor of type $(1,3)$, called the **Riemann curvature tensor** (AKA Riemann tensor, curvature tensor, Riemann-Christoffel tensor):

$$R^c{}_{dab}u^a v^b w^d \equiv u^a \nabla_a \left(v^b \nabla_b w^c\right) - v^b \nabla_b \left(u^a \nabla_a w^c\right) - [u,v]^d \nabla_d w^c$$
$$= u^a v^b \nabla_a \nabla_b w^c - u^a v^b \nabla_b \nabla_a w^c + T^d{}_{ab} u^a v^b \nabla_d w^c$$
$$\Rightarrow R^c{}_{dab} w^d = \left(\nabla_a \nabla_b - \nabla_b \nabla_a + T^d{}_{ab} \nabla_d\right) w^c$$

Here we have used the Leibniz rule and recalled that $[u,v]^d = u^a \nabla_a v^d - v^b \nabla_b u^d - T^d{}_{ab} u^a v^b$.

To obtain an expression in terms of the connection coefficients, we first examine the double covariant derivative, recalling that $\nabla_b w^c$ is a tensor:

$$\nabla_a \left(\nabla_b w^c\right) = \partial_a \nabla_b w^c + \Gamma^c{}_{fa} \nabla_b w^f - \Gamma^f{}_{ba} \nabla_f w^c$$
$$= \partial_a \partial_b w^c + \partial_a \left(\Gamma^c{}_{fb} w^f\right)$$
$$\quad + \Gamma^c{}_{fa} \partial_b w^f + \Gamma^c{}_{fa} \Gamma^f{}_{gb} w^g - \Gamma^f{}_{ba} \nabla_f w^c$$
$$= \partial_a \partial_b w^c + \partial_a \Gamma^c{}_{fb} w^f$$
$$\quad + \Gamma^c{}_{fb} \partial_a w^f + \Gamma^c{}_{fa} \partial_b w^f$$
$$\quad + \Gamma^c{}_{fa} \Gamma^f{}_{gb} w^g - \Gamma^f{}_{ba} \nabla_f w^c.$$

When we subtract the same expression with a and b reversed, we recognize that for the functions w^c we have $\partial_a \partial_b w^c - \partial_b \partial_a w^c = [e_a, e_b]^d \partial_d w^c$, that the second line $\Gamma^c{}_{fb} \partial_a w^f + \Gamma^c{}_{fa} \partial_b w^f$ vanishes, and that $\Gamma^f{}_{ba} - \Gamma^f{}_{ab} = [e_a, e_b]^f + T^f{}_{ab}$, so that

$$\left(\nabla_a \nabla_b - \nabla_b \nabla_a \right) w^c = [e_a, e_b]^d \partial_d w^c + \partial_a \Gamma^c{}_{fb} w^f - \partial_b \Gamma^c{}_{fa} w^f$$
$$+ \Gamma^c{}_{fa} \Gamma^f{}_{gb} w^g - \Gamma^c{}_{fb} \Gamma^f{}_{ga} w^g$$
$$- \left([e_a, e_b]^f + T^f{}_{ab} \right) \nabla_f w^c,$$

and thus relabeling dummy indices to obtain an expression in terms of w^d, we arrive at

$$R^c{}_{dab} w^d = \left(\nabla_a \nabla_b - \nabla_b \nabla_a + T^d{}_{ab} \nabla_d \right) w^c$$
$$= \left(\partial_a \Gamma^c{}_{db} - \partial_b \Gamma^c{}_{da} + \Gamma^c{}_{fa} \Gamma^f{}_{db} - \Gamma^c{}_{fb} \Gamma^f{}_{da} - [e_a, e_b]^f \Gamma^c{}_{df} \right) w^d.$$

This expression follows much more directly from the expression $\check{R} \equiv \mathrm{d}\check{\Gamma} + \check{\Gamma} \wedge \check{\Gamma}$, but the above derivation from the covariant derivative expression is included here to clarify other presentations which are sometimes obscured by the quirks of index notation for covariant derivatives.

$\triangle$ The derivation above makes clear how the expression for the curvature in terms of the covariant derivative simplifies to $R^c{}_{dab} w^d = \left(\nabla_a \nabla_b - \nabla_b \nabla_a \right) w^c$ for zero torsion but is unchanged in a holonomic frame, while in contrast the expression in terms of the connection coefficients is unchanged for zero torsion but in a holonomic frame simplifies to omit the term $[e_a, e_b]^f \Gamma^c{}_{df} w^d$.

$\triangle$ Note that the sign and the order of indices of R as a tensor are not at all consistent across the literature.

9.2.6 *First Bianchi identity*

If we take the exterior covariant derivative of the torsion, we get

$$\mathrm{D}\vec{T} = \mathrm{DD}\vec{\beta} = \check{R} \wedge \vec{\beta}.$$

This is called the **first** (AKA algebraic) **Bianchi identity**. Using the antisymmetry of $\check{R}$, we can write the first Bianchi identity explicitly as

$$\mathrm{D}\vec{T}(u,v,w) = \check{R}(u,v)\vec{w} + \check{R}(v,w)\vec{u} + \check{R}(w,u)\vec{v}.$$

In the case of zero torsion, this identity becomes $\check{R} \wedge \vec{\beta} = 0$, which in index notation can be written $R^c{}_{[dab]} = 0$.

We can find a geometric interpretation for this identity by first constructing a variant of our picture of $\check{R}(u,v)\vec{w}$ as the change in $\vec{w}$ after being parallel transported in opposite directions around a loop. Taking advantage of our previous result that $\check{R}(u,v)\vec{w}$ only depends upon the local values of u and v, we are free to construct their vector field values such that $[u,v] = 0$. We then examine the difference between $\vec{w}$ being parallel transported in each direction halfway around the loop. For infinitesimal parallel transport from a point p along a curve C with tangent v we have $\|_{\varepsilon v}(w|_p) \equiv \|_C(w|_p) = w|_{p+\varepsilon v} - \varepsilon \nabla_v w|_p$. Therefore we find that

$$\|_{\varepsilon u}\|_{\varepsilon v}(w|_p) = \|_{\varepsilon u}(w|_{p+\varepsilon v} - \varepsilon\nabla_v w|_p)$$
$$= w|_{p+\varepsilon v+\varepsilon u} - \varepsilon\nabla_v w|_{p+\varepsilon u} - \varepsilon\nabla_u w|_{p+\varepsilon v} + \varepsilon^2\nabla_u\nabla_v w|_p,$$

so that

$$\|_{\varepsilon u}\|_{\varepsilon v}(w|_p) - \|_{\varepsilon v}\|_{\varepsilon u}(w|_p) = \varepsilon^2\nabla_u\nabla_v w|_p - \varepsilon^2\nabla_v\nabla_u w|_p$$
$$= \varepsilon^2\check{R}(u,v)\vec{w},$$

since $[u,v] = 0$ means that $w|_{p+\varepsilon v+\varepsilon u} = w|_{p+\varepsilon u+\varepsilon v}$. In the case of zero torsion, we can further take advantage of our freedom in choosing the vector field values of u and v by requiring them to equal their parallel transports, i.e. $v|_{p+\varepsilon u} \equiv \|_{\varepsilon u}(v|_p)$ and $u|_{p+\varepsilon v} \equiv \|_{\varepsilon v}(u|_p)$, preserving the property $[u,v] = 0$ due to the vanishing torsion.

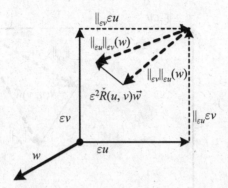

Figure 9.2.7 A slight variant of $\check{R}(u, v)\,\vec{w}$ viewed as "the difference between w when parallel transported around the two opposite edges of the boundary of the surface defined by its arguments." In the case of zero torsion, the boundary can be built from parallel transports instead of vector field values.

Thus, still assuming zero torsion, we can construct a cube from the parallel translations of u, v, and w. This construction reveals that the first Bianchi identity corresponds to the fact that the three curvature vectors form a triangle, i.e. their sum is zero.

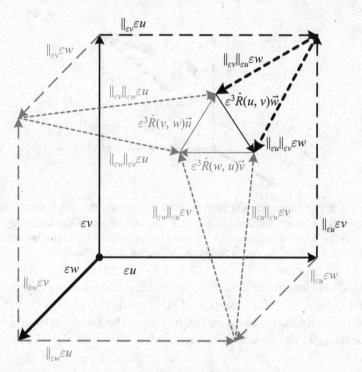

Figure 9.2.8 The first Bianchi identity reflects the fact that for zero torsion, the far corners of a cube made of parallel transported vectors do not meet, and their separation is made up of the differences in parallel transport via opposite edges of each face. Note that the corners of the triangle are points since vanishing torsion means that e.g. $\varepsilon u + \|_{\varepsilon u} (\varepsilon w) = \varepsilon w + \|_{\varepsilon w} (\varepsilon u)$, so that the top point of the triangle reflects this equality parallel transported by εv.

9.2.7 Second Bianchi identity

If we take the exterior covariant derivative of the curvature, we get

$$D\check{R} = 0.$$

This is called the **second Bianchi identity**, and can be verified algebraically from the definition $\check{R} \equiv d\check{\Gamma} + \check{\Gamma} \wedge \check{\Gamma}$. We can write this identity more explicitly as

$$
\begin{aligned}
0 &= D\check{R}(u, v, w)\vec{a} \\
&= \nabla_u \check{R}(v, w)\vec{a} + \nabla_v \check{R}(w, u)\vec{a} + \nabla_w \check{R}(u, v)\vec{a} \\
&\quad - \check{R}([u, v], w)\vec{a} - \check{R}([v, w], u)\vec{a} - \check{R}([w, u], v)\vec{a},
\end{aligned}
$$

where we have used the antisymmetry of $\check{R}$ and the covariant derivative acts on the value of $\check{R}$ as a tensor of type $(1,1)$. Working this expression into tensor notation and using the tensor expression for the torsion in terms of the commutator, we find that

$$0 = \nabla_e R^c{}_{dab} + \nabla_a R^c{}_{dbe} + \nabla_b R^c{}_{dea}$$
$$- R^c{}_{dfe} T^f{}_{ab} - R^c{}_{dfa} T^f{}_{be} - R^c{}_{dfb} T^f{}_{ea},$$

or

$$R^c{}_{d[ab;e]} = R^c{}_{df[e} T^f{}_{ab]},$$

and in the case of zero torsion, $R^c{}_{d[ab;e]} = 0$.

Geometrically, the second Bianchi identity can be seen as reflecting the same "boundary of a boundary" idea as that of $\mathrm{d}^2 = 0$ in Figure 6.3.8, except that here we are parallel transporting a vector $\vec{a}$ around each face that makes up the boundary of the cube. As in the previous section, we can take advantage of the fact that $\check{R}(v,w)\vec{a}$ only depends upon the local value of $\vec{a}$, constructing its vector field values such that e.g. $\vec{a}|_{p+\varepsilon u} = \|_{\varepsilon u}(\vec{a}|_p)$, giving us

$$\varepsilon \nabla_u \check{R}(v,w)\vec{a} = \check{R}(v|_{p+\varepsilon u}, w|_{p+\varepsilon u})\vec{a}|_{p+\varepsilon u} - \|_{\varepsilon u}\check{R}(v,w)\|_{\varepsilon u}^{-1}\vec{a}|_{p+\varepsilon u}$$
$$= \check{R}(v|_{p+\varepsilon u}, w|_{p+\varepsilon u})\|_{\varepsilon u}\vec{a} - \|_{\varepsilon u}\check{R}(v,w)\vec{a}.$$

The first term parallel translates $\vec{a}$ along εu and then around the parallelogram defined by v and w at $p + \varepsilon u$, while the second parallel translates $\vec{a}$ around the parallelogram defined by v and w at p, then along εu. Thus in the case of vanishing Lie commutators (e.g. a holonomic frame), we construct a cube from the vector fields u, v, and w, and find that the second Bianchi identity reflects the fact that $\mathrm{D}\check{R}(u,v,w)\vec{a}$ parallel translates $\vec{a}$ along each edge of the cube an equal number of times in opposite directions, thus canceling out any changes.

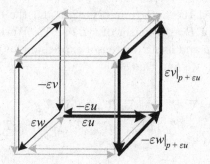

Figure 9.2.9 The second Bianchi identity reflects the fact that for vanishing Lie commutators, $\mathrm{D}\check{R}(u,v,w)\vec{a}$ parallel translates $\vec{a}$ along each edge of the cube made of the three vector field arguments an equal number of times in opposite directions, thus canceling out any changes. Above, $\varepsilon\nabla_u\check{R}(v,w)\vec{a} = \check{R}(v|_{p+\varepsilon u}, w|_{p+\varepsilon u})\,\|_{\varepsilon u}\,\vec{a} - \|_{\varepsilon u}\check{R}(v,w)\vec{a}$ is highlighted by the bold arrows representing the path along which $\vec{a}$ is parallel translated in the first term, and by the remaining dark arrows representing the path along which $\vec{a}$ is parallel translated in the second term.

In the case of a non-vanishing commutator, e.g. $[u,v] \neq 0$, we find that the cube gains a "shaved edge," and that the extra non-vanishing term $-\check{R}([u,v],w)\vec{a}$ in $\mathrm{D}\check{R}$ maintains the "boundary of a boundary" logic by adding a loop of parallel translation of $\vec{a}$ in the proper direction around the new "face" created.

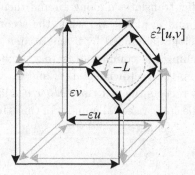

Figure 9.2.10 In the case of a non-vanishing commutator, the extra term $-\check{R}([u,v],w)\vec{a}$ in $\mathrm{D}\check{R}$ maintains the cancellation of face boundaries by adding a loop L around the new "shaved edge" created.

9.2.8   The holonomy group

We have seen that on a manifold with connection, the curvature measures the effect of a vector being parallel transported around an infinitesimal loop. If we consider the set of all closed loops at a basepoint p on a connected manifold with connection M^n, the associated linear transformations due to parallel transport of a vector around each loop form a group called the **holonomy group** $\text{Hol}(M)$. The **restricted holonomy group** $\text{Hol}^0(M)$ only counts loops homotopic to zero.

From the definition of the parallel transporter we can see that $\text{Hol}(M)$ is in fact a group, and also a subgroup of $GL(\mathbb{R}^n)$, and therefore a Lie group. We can also see that for a connected manifold it is independent of the basepoint p, since changing basepoints induces a similarity transformation (change of basis), altering the matrix representation of the group but acting as an isomorphism on the abstract group. If M is simply connected, then $\text{Hol}^0(M) = \text{Hol}(M)$; if not, then $\text{Hol}^0(M)$ is the identity component of $\text{Hol}(M)$, which is a group representation of the fundamental group of M called the **monodromy representation**.

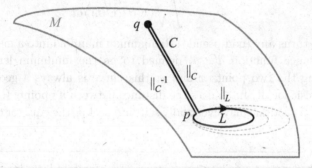

Figure 9.2.11 The holonomy group is comprised of elements $\|_L \in GL(\mathbb{R}^n)$ associated with parallel translation around loops; changing the basepoint of the loop from p to q induces a similarity translation $\|_C^{-1}\|_L\|_C$, leaving the abstract group unchanged.

Since the curvature is the infinitesimal version of the holonomy construction, we might expect that it be related to the Lie algebra of $\text{Hol}(M)$, which is called the **holonomy algebra**. The **Ambrose-Singer theorem** confirms this; in the case of a simply connected manifold, it says that the Lie algebra of $\text{Hol}(M)$ is generated by all elements of the form $\check{R}(v, w)\,|_{q \in M}$.

Zero holonomy group then implies zero curvature; but the converse is

only true for the restricted holonomy group, as can be seen by considering
e.g. a flat sheet of paper rolled into a cone. However, zero curvature implies
that the holonomy algebra vanishes, which means that the holonomy group
is discrete.

9.3 Introducing lengths and angles

9.3.1 *The Riemannian metric*

Recall from Section 3.2.4 that a (pseudo) metric tensor is a (pseudo) inner
product $\langle v, w \rangle$ on a vector space V that can be represented by a symmetric
tensor g_{ab}, and thus can be used to lower and raise indices on tensors. A
(pseudo) Riemannian metric (AKA metric) is a (pseudo) metric tensor
field on a manifold M, making M a **(pseudo) Riemannian manifold**.

A metric defines the length (norm) of tangent vectors, and can thus be
used to define the length L of a curve C via parametrization and integration:

$$L(C) \equiv \int \left\| \dot{C}(t) \right\| \mathrm{d}t$$

$$= \int \sqrt{\left\langle \dot{C}(t), \dot{C}(t) \right\rangle} \mathrm{d}t$$

This also turns any (non-pseudo) Riemannian manifold into a metric space,
with distance function $d(x, y)$ defined to be the minimum length curve
connecting the two points x and y; this curve is always a geodesic, and
any geodesic locally minimizes the distance between its points (only locally
since e.g. a geodesic may eventually self-intersect as the equator on a sphere
does).

✿ With a metric, our intuitive picture of a manifold loses its "stretchi-
ness" via the introduction of length and angles; but having only intrinsically defined properties, the manifold can still be e.g. rolled up like
a piece of paper if imagined as flat and embedded in a larger space.

If the coordinate frame of x^μ is orthonormal at a point $p \in M^n$ in
a Riemannian manifold, for arbitrary coordinates y^μ we can consider the

components of the metric tensor in the two coordinate frames to find that

$$g_{\mu\nu}\mathrm{d}y^{\mu}\mathrm{d}y^{\nu} = \delta_{\lambda\sigma}\mathrm{d}x^{\lambda}\mathrm{d}x^{\sigma}$$

$$= \delta_{\lambda\sigma}\frac{\partial x^{\lambda}}{\partial y^{\mu}}\mathrm{d}y^{\mu}\frac{\partial x^{\sigma}}{\partial y^{\nu}}\mathrm{d}y^{\nu}$$

$$= \left[J_x(y)\right]^{T}\left[J_x(y)\right]\mathrm{d}y^{\mu}\mathrm{d}y^{\nu}$$

$$\Rightarrow \det\left(g_{\mu\nu}\right) = \left[\det\left(J_x\left(y\right)\right)\right]^{2},$$

where $J_x(y)$ is the Jacobian matrix and we have used the fact that $\det(A^T A) = [\det(A)]^2$. Thus the volume of an region $U \in M^n$ corresponding to $R \in \mathbb{R}^n$ in the coordinates x^μ is

$$V(U) = \int_R \sqrt{\det(g)}\mathrm{d}x^1\dots\mathrm{d}x^n,$$

where $\det(g)$ is the determinant of the metric tensor as a matrix in the coordinate frame $\partial/\partial x^\mu$. In the context of a pseudo-Riemannian manifold $\det(g)$ can be negative, and the integrand $\mathrm{d}V \equiv \sqrt{|\det(g)|}\mathrm{d}x^1\dots\mathrm{d}x^n$ is called the **volume element**, or when written as a form $\mathrm{d}V \equiv \sqrt{|\det(g)|}\mathrm{d}x^1 \wedge \dots \wedge \mathrm{d}x^n$ it is called the **volume form**. In physical applications $\mathrm{d}V$ usually denotes the **volume pseudo-form**, which gives a positive value regardless of orientation. Note that if the coordinate frame is orthonormal then $|\det(g)| = 1$; thus these definitions are consistent with those previously defined. Sometimes one defines a volume form on a manifold without defining a metric; in this case the metric (and connection) is not uniquely determined.

△ The symbol g is frequently used to denote $\det(g)$, and sometimes $\sqrt{|\det(g)|}$, in addition to denoting the metric tensor itself.

We can use the inner product to define an **orthonormal frame** on M. In four dimensions an orthonormal frame is also called a **tetrad** (AKA vierbein). Any frame on a manifold can be defined to be an orthonormal frame, which is equivalent to defining the metric (which in the orthonormal frame is $g_{ab} = \eta_{ab}$). An orthonormal holonomic frame exists on a region of M if and only if that region is flat. Thus in general, given a set of coordinates on M, we have to choose between using either a non-coordinate orthonormal frame or a non-orthonormal coordinate frame.

At a point $p \in U \subset M$, an orthonormal basis for T_pU can be used to form geodesic normal coordinates, which are then called **Riemann normal coordinates**. In these coordinates the partial derivatives of the metric

$g_{ab} = \eta_{ab}$ all vanish at p. The **Hopf-Rinow theorem** says that a connected
Riemannian manifold M is complete as a metric space (or equivalently, all
closed and bounded subsets are compact) if and only if it is **geodesically
complete**, meaning that the exponential map is defined for all vectors at
some $p \in M$. If M is geodesically complete at p, then it is at all points
on the manifold, so this property can also be used to state the theorem.
This theorem is not valid for pseudo-Riemannian manifolds; any (pseudo)
Riemannian manifold that is geodesically complete is called a **geodesic
manifold**.

As noted previously, a Riemannian metric can be defined on any differentiable manifold. In general, however, not every manifold admits a
pseudo-Riemannian metric, and in particular not every 4-manifold admits
a Minkowski metric; but 4-manifolds that are noncompact, parallelizable,
or compact, connected and of Euler characteristic 0 all do.

In the same way that differentiable manifolds are equivalent if they are
related by a diffeomorphism, Riemannian manifolds are equivalent if they
are related by an **isometry**, a diffeomorphism $\Phi \colon M \to N$ that preserves
the metric, i.e. $\forall v, w \in TM$, $\langle v, w \rangle |_p = \langle d\Phi_p(v), d\Phi_p(w) \rangle |_{\Phi(p)}$. Also like
diffeomorphisms, the isometries of a manifold form a group; for example, the
group of isometries of Minkowski space is the Poincaré group. A vector field
whose one-parameter diffeomorphisms are isometries is called a **Killing
field**, also called a **Killing vector** since it can be shown (Petersen [2006]
pp. 188-189) that a Killing field is determined by a vector at a single point
along with its covariant derivatives. A Killing field thus satisfies $L_v g_{ab} =
0$, which for a Levi-Civita connection (see next section) is equivalent to
$\nabla_a v_b + \nabla_b v_a = 0$, called the **Killing equation** (AKA Killing condition).

We can then consider isometric immersions and embeddings, and ask
whether every Riemannian manifold can be embedded in some $\mathbb{R}^n$. The
Nash embedding theorem provides an affirmative answer, and it can
also be shown that every pseudo-Riemanian manifold can be isometrically
embedded in some $\mathbb{R}^n$ with some signature while maintaining arbitrary
differentiability of the metric.

9.3.2 The Levi-Civita connection

A connection on a Riemannian manifold M is called a **metric connection**
(AKA metric compatible connection, isometric connection) if its associated
parallel transport respects the metric, i.e. it preserves lengths and angles.
More precisely, $\forall v, w \in TM$, we require that $\langle \|_C (v), \|_C (w) \rangle = \langle v, w \rangle$ for

any curve C in M. This means that the holonomy group is a subgroup of $O(n)$, or of $SO(n)$ if (and only if) M is orientable.

In terms of the metric, this can be written $g_{ab} \|_C v^a \|_C w^b = g_{ab} v^a w^b$. Then since $\nabla_c \langle v, w \rangle = 0$, and recalling that the covariant derivative on the tensor algebra was defined to respect parallel translation, we also have for infinitesimal C that $\|_C (g_{ab} v^a w^b) = \|_C g_{ab} \|_C v^a \|_C w^b = g_{ab} v^a w^b$, so that we must have $\|_C g_{ab} = g_{ab}$, or $\nabla_c g_{ab} = 0$. In terms of the connection coefficients, a metric connection then satisfies

$$\nabla_c g_{ab} = \partial_c g_{ab} - \Gamma^d{}_{ac} g_{db} - \Gamma^d{}_{bc} g_{ad} = 0.$$

Using the Leibniz rule for the covariant derivative over the tensor product, we can derive a Leibniz rule over the inner product:

$$\nabla_c \left(g_{ab} v^a w^b \right) = 0 + g_{ab} \nabla_c v^a w^b + g_{ab} v^a \nabla_c w^b$$
$$\Rightarrow \nabla_u \langle v, w \rangle = \langle \nabla_u v, w \rangle + \langle v, \nabla_u w \rangle$$

Requiring this relationship to hold is an equivalent way to define a metric connection.

The **Levi-Civita connection** (AKA Riemannian connection, Christoffel connection) is then the torsion-free metric connection on a (pseudo) Riemannian manifold M. The **fundamental theorem of Riemannian geometry** states that for any (pseudo) Riemannian manifold the Levi-Civita connection exists and is unique. On the other hand, an arbitrary connection can only be the Levi-Civita connection for some metric if it is torsion-free and preserves lengths. More precisely, given a simply connected manifold M with a torsion-free connection, a metric of signature (r, s) compatible with this connection exists if and only if $\text{Hol}(M) \subseteq O(r, s)$; moreover, this metric is unique only up to a scaling factor (in physics, this corresponds to a choice of units).

For a metric connection, the curvature then must take values that are infinitesimal rotations, i.e. $\check{R}$ is $o(r, s)$-valued. Thus if we eliminate the influence of the signature by lowering the first index, the first two indices of the curvature tensor are anti-symmetric:

$$R_{cdab} = -R_{dcab}$$

Using the anti-symmetry of the other indices and the first Bianchi identity, this leads to another commonly noted symmetry

$$R_{cdab} = R_{abcd}.$$

The Leibniz rule for the covariant derivative over the inner product can be used to derive an expression called the **Koszul formula**:

$$2 \langle \nabla_u v, w \rangle = \nabla_u \langle v, w \rangle + \nabla_v \langle w, u \rangle - \nabla_w \langle u, v \rangle$$
$$- \langle u, [v, w] \rangle + \langle v, [w, u] \rangle + \langle w, [u, v] \rangle$$

Substituting in the frame vector fields and eliminating the metric tensor from the left hand side, we arrive at an expression for the connection in terms of the metric:

$$2\Gamma^c{}_{ba} = g^{cd}(\partial_a g_{bd} + \partial_b g_{da} - \partial_d g_{ab}$$
$$- g_{af}[e_b, e_d]^f + g_{bf}[e_d, e_a]^f + g_{df}[e_a, e_b]^f)$$

On a Riemannian manifold, the connection coefficients for the Levi-Civita connection in a coordinate basis $\Gamma^\lambda{}_{\mu\sigma}$ are called the **Christoffel symbols**. Thus the Christoffel symbols are determined by the partial derivatives of the metric, which means that the Christoffel symbols vanish at the origin of Riemann normal coordinates.

☆ The vanishing of the Christoffel symbols at the origin of Riemann normal coordinates is frequently used to simplify the derivation of tensor relations which are then, being frame-independent, seen to be true in any coordinate system or frame (and if the origin was chosen arbitrarily, at any point). In particular, the covariant and partial derivatives are equivalent at the origin of Riemann normal coordinates.

9.3.3 *Independent quantities and dependencies*

While in general the curvature on a Riemannian manifold does not determine the metric, for a manifold with connection that is compact, simply connected, and has no regions of constant curvature (i.e. there is no way to "stretch" the manifold without affecting the curvature), knowledge of the curvature at all points determines the connection (up to changes in frame), and therefore the metric that makes this connection Levi-Civita (up to a constant scaling factor).

If we choose coordinate charts and use coordinate frames on M^n, we can calculate the number of independent functions and equations associated with the various quantities and relations we have covered, and use them to verify the associated dependencies.

Table 9.3.1 Independent function and equation counts in a coordinate frame.

Quantity / relation	Viewpoint	Count
Metric	Symmetric matrix of functions	$n(n+1)/2$
Coordinate frame	Fixed	0
Connection	$gl(n, \mathbb{R})$-valued 1-form	n^3
Metric condition	Derivative of metric	$n^2(n+1)/2$
Torsion-free condition	$\mathbb{R}^n$-valued 2-form	$n^2(n-1)/2$

The choice of coordinates determines the frame, leaving the geometry of the Riemannian manifold defined by the $n(n+1)/2$ functions of the metric. A torsion-free connection consists of $n^3 - n^2(n-1)/2 = n^2(n+1)/2$ functions. The metric condition is exactly this number of equations, allowing us in general to solve for the connection if the metric is known, or vice-versa (up to a constant scaling factor).

Alternatively, we can look at things in a orthonormal frame:

Table 9.3.2 Independent function and equation counts in an orthonormal frame.

Quantity / relation	Viewpoint	Count
Metric	Fixed	0
Orthonormal frame	n vector fields	n^2
Change of orthonormal frame	$SO(n)$-valued 0-form	$n(n-1)/2$
Connection	$so(n)$-valued 1-form	$n^2(n-1)/2$
Metric condition	Automatically satisfied	0
Torsion-free condition	$\mathbb{R}^n$-valued 2-form	$n^2(n-1)/2$

Here the metric is constant, and the orthonormal frame consists of n^2 functions, but it is determined only up to a change of orthonormal frame (rotation), leaving $n^2 - n(n-1)/2 = n(n+1)/2$ functions, consistent with the metric function count above. The torsion-free condition is the same number of equations as the connection has functions, so that in general the torsion-free connection can be determined by the orthonormal frame.

9.3.4 *The divergence and conserved quantities*

Recall from Section 6.3.5 that the divergence of a vector field u can be generalized to a pseudo-Riemannian manifold (sometimes called the **covariant divergence**) by defining $\mathrm{div}(u) \equiv *\mathrm{d}(*(u^\flat))$. Using the previously stated relations $i_u\Omega = *(u^\flat)$ and $A = (*A)\Omega$ for $A \in \Lambda^n M^n$, we have $d(i_u\Omega) = d(*(u^\flat)) = *d(*(u^\flat))\Omega = \text{div}(u)\Omega$. Using $i_u d + d i_u = L_u$ we then arrive at $\text{div}(u)\Omega = L_u\Omega$, or as it is more commonly written

$$\text{div}(u)dV = L_u dV.$$

Thus we can say that $\text{div}(u)$ is "the fraction by which a unit volume changes when transported by the flow of u," and if $\text{div}(u) = 0$ then we can say that "the flow of u leaves volumes unchanged." Expanding the volume element in coordinates x^λ we can obtain an expression for the divergence in terms of these coordinates,

$$\text{div}(u) = \frac{1}{\sqrt{|\det(g)|}}\partial_\lambda \left(u^\lambda \sqrt{|\det(g)|} \right).$$

Note that both this expression and $\nabla_a u^a$ are coordinate-independent and equal to $\partial_a u^a$ in Riemann normal coordinates, confirming our expectation that in general we have

$$\text{div}(u) = \nabla_a u^a.$$

Using the relation $\text{div}(u)\Omega = d(i_u\Omega)$ above, along with Stokes' theorem, we recover the classical **divergence theorem**

$$\int_V \text{div}(u)dV = \int_{\partial V} i_u dV$$
$$= \int_{\partial V} \langle u, \hat{n} \rangle \, dS,$$

where V is an n-dimensional compact submanifold of M^n, $\hat{n}$ is the unit normal vector to ∂V, and $dS \equiv i_{\hat{n}} dV$ is the induced volume element ("surface element") for ∂V. In the case of a Riemannian metric, this can be thought of as reflecting the intuitive fact that "the change in a volume due to the flow of u is equal to the net flow across that volume's boundary." If $\text{div}(u) = 0$ then we can say that "the net flow of u across the boundary of a volume is zero." We can also consider an infinitesimal V, so that the divergence at a point measures "the net flow of u across the boundary of an infinitesimal volume." As usual, for a pseudo-Riemannian metric these geometric intuitions have less meaning.

The divergence can be extended to contravariant tensors T by defining $\text{div}(T) \equiv \nabla_a T^{ab}$, although other conventions are in use. Since $\text{div}(T)$ is vector-valued and the parallel transport of vectors is path-dependent, we cannot in general integrate to get a divergence theorem for tensors. In the case of a flat metric however, we are able to integrate to get a divergence theorem for each component

$$\int_V \nabla_a T^{ab} dV = \int_{\partial V} T_a{}^b \hat{n}^a dS.$$

In physics, the vector field u often represents the **current vector** (AKA current density, flux, flux density) $j \equiv \rho u$ of an actual physical flow, where ρ is the density of the physical quantity Q and u is thus a velocity field; e.g. in $\mathbb{R}^3$, j has units $Q/(\text{length})^2(\text{time})$. There are several quantities that can be defined around this concept:

Table 9.3.3 Quantities related to current.

Quantity	Definition	Meaning
Current vector	$j \equiv \rho u$	The vector whose length is the amount of Q per unit time crossing a unit area perpendicular to j
Current form	$\zeta \equiv i_j dV$ $= \langle j, \hat{n} \rangle\, dS$	The $(n-1)$-form which gives the amount of Q per unit time crossing the area defined by the argument vectors
Current density	$\mathsf{j} \equiv \sqrt{|\det(g)|}\, j$ $\Rightarrow \zeta = \langle \mathsf{j}, \hat{n} \rangle\, dx^{\lambda_1} \wedge \cdots \wedge dx^{\lambda_{n-1}}$	The vector whose length is the amount of Q per unit time crossing a unit coordinate area perpendicular to j
Current	$I \equiv \int_S \zeta$ $= \int_S \langle j, \hat{n} \rangle\, dS$ $= \int_{S(x^\lambda)} \langle \mathsf{j}, \hat{n} \rangle\, dx^{\lambda_1} \cdots dx^{\lambda_{n-1}}$	The amount of Q per unit time crossing S
Current 4-vector	$J \equiv (\rho, j^\mu)$	Current vector on the spacetime manifold

Notes: ρ is the density of the physical quantity Q, u is a velocity field, $\hat{n}$ is the unit normal to a surface S, and x^λ are coordinates on the submanifold S. The current 4-vector can be generalized to other Lorentzian manifolds, and can also be turned into a form or a density.

$\triangle$ Note that the terms flux and current (as well as flux density and current density) are not used consistently in the literature.

The current density j is an example of a **tensor density**, which in general takes the form $\mathfrak{T} \equiv \left(\sqrt{|\det(g)|} \right)^{W} T$, where T is a tensor and W is called the **weight**. Note that tensor densities are not coordinate-independent quantities.

For a Riemannian metric we now define the **continuity equation** (AKA equation of continuity)

$$\frac{dq}{dt} = \Sigma - \int_{\partial V} \langle j, \hat{n} \rangle \, dS,$$

where q is the amount of Q contained in V, t is time, and Σ is the rate of Q being created within V. The continuity equation thus states the intuitive fact that the change of Q within V equals the amount generated less the amount which passes through ∂V. Using the divergence theorem, we can then obtain the differential form of the continuity equation

$$\frac{\partial \rho}{\partial t} = \sigma - \text{div}(j),$$

where σ is the amount of Q generated per unit volume per unit time. This equation then states the intuitive fact that at a point, the change in density of Q equals the amount generated less the amount that moves away. Positive σ is referred to as a **source** of Q, and negative σ a **sink**. If $\sigma = 0$ then we say that Q is a **conserved quantity** and refer to the continuity equation as a (local) **conservation law**.

Under a flat Lorentzian metric, we can combine ρ and j into the current 4-vector J and express the continuity equation with $\sigma = 0$ as

$$\text{div}(J) = 0,$$

whereupon J is called a **conserved current**. Note that in this approach we lose the intuitive meaning of the divergence under a Riemannian metric. If any curvature is present, when we split out the time component we recover a Riemannian divergence but introduce a source due to the non-zero Christoffel symbols

$$\nabla_{\mu} J^{\mu} = \partial_{\mu} J^{\mu} + \Gamma^{\mu}{}_{\nu\mu} J^{\nu}$$
$$= \partial_{t} \rho + \nabla_{i} j^{i} + \left(\Gamma^{\mu}{}_{t\mu} \rho + \Gamma^{t}{}_{it} j^{i} \right),$$

where t is the negative signature component and the index i goes over the remaining positive signature components. Thus, since the Christoffel symbols are coordinate-dependent, in the presence of curvature there is in general no coordinate-independent conserved quantity associated with a vanishing Lorentzian divergence.

Several methodologies can be used to derive conserved quantities and currents from an expression that in some way describes a physical system (and is often call simply "the system"); in particular, **Noether's theorem** derives conserved currents from transformations ("symmetries") on the variables of an expression called the **action** that leave it unchanged.

9.3.5 *Ricci and sectional curvature*

The **Ricci curvature tensor** (AKA Ricci tensor) is formed by contracting two indices in the Riemann curvature tensor:

$$R_{ab} \equiv R^c{}_{acb}$$
$$\mathrm{Ric}(v, w) \equiv R_{ab} v^a w^b$$

Using the symmetries of the Riemann tensor for a metric connection along with the first Bianchi identity for zero torsion, it is easily shown that the Ricci tensor is symmetric. A pseudo-Riemannian manifold is said to have constant Ricci curvature, or to be an **Einstein manifold**, if the Ricci tensor is a constant multiple of the metric tensor.

Since the Ricci tensor is symmetric, by the spectral theorem it can be diagonalized and thus is determined by

$$\mathrm{Ric}(v) \equiv \mathrm{Ric}(v, v),$$

which is called the **Ricci curvature function** (AKA Ricci function). Note that the Ricci function is not a 1-form since it is not linear in v. Choosing a basis that diagonalizes R_{ab} is equivalent to choosing our basis vectors to line up with the directions that yield extremal values of the Ricci function on the unit vectors $\mathrm{Ric}(\hat{v}, \hat{v})$ (or equivalently, the principal axes of the ellipsoid / hyperboloid $\mathrm{Ric}(v, v) = 1$).

Finally, if we raise one of the indices of the Ricci tensor and contract we arrive at the **Ricci scalar** (AKA scalar curvature):

$$R \equiv g^{ab} R_{ab}$$

For a Riemannian manifold M^n, the Ricci scalar can thus be viewed as n times the average of the Ricci function on the set of unit tangent vectors.

$\triangle$ The Ricci function and Ricci scalar are sometimes defined as averages instead of contractions (sums), introducing extra factors in terms of the dimension n to the above definitions.

The Ricci function in terms of the curvature 2-form in an orthonormal frame e_μ (dropping the hats to avoid clutter) on a pseudo-Riemannian manifold M^n naturally splits into terms which each also measure curvature:

$$\mathrm{Ric}(e_\mu) = \sum_{i \neq \mu} g_{ii} \left\langle \check{R}(e_i, e_\mu)\vec{e}_\mu, e_i \right\rangle$$

The term $i = \mu$ vanishes due to the anti-symmetry of $\check{R}$. The $(n-1)$ non-zero terms are each called a **sectional curvature**, which in general is defined as

$$K(v, w) \equiv \frac{\left\langle \check{R}(v, w)\vec{w}, v \right\rangle}{\langle v, v \rangle \langle w, w \rangle - \langle v, w \rangle^2}$$

$$\Rightarrow K(e_i, e_j) = g_{ii}g_{jj} \left\langle \check{R}(e_i, e_j)\vec{e}_j, e_i \right\rangle$$

$$\Rightarrow \mathrm{Ric}(e_\mu) = \sum_{i \neq \mu} g_{\mu\mu} K(e_i, e_\mu)$$

$$\Rightarrow R = \sum_j g_{jj} \mathrm{Ric}(e_j)$$

$$= \sum_{i \neq j} K(e_i, e_j)$$

$$= 2 \sum_{i < j} K(e_i, e_j).$$

Note that the sectional curvature is not a 2-form since it is not linear in its arguments; in fact it is constructed to only depend on the plane defined by them, and therefore is symmetric and vanishes for equal arguments. Thus for a Riemannian manifold, the Ricci function of a unit vector $\mathrm{Ric}(\hat{v})$ can be viewed as $(n-1)$ times the average of the sectional curvatures of the planes that include $\hat{v}$, and the Ricci scalar can be viewed as n times the average of all the Ricci functions. For a pseudo-Riemannian manifold, the Ricci scalar is twice the sum of all sectional curvatures, or $n(n-1)$ times the average of all sectional curvatures, whose count is the binomial coefficient n choose 2 or $n(n-1)/2$.

The **Cartan-Hadamard theorem** states that the universal covering space of a complete Riemannian manifold M^n with non-positive sectional curvature is diffeomorphic to $\mathbb{R}^n$. For M^n complete with constant sectional curvature K (sometimes called a **space form**), its universal covering space is isometric to $\mathbb{R}^n$ if $K = 0$, S^n if $K = 1$, and H^n if $K = -1$, where H^n is called the (real) **hyperbolic space** and we can generalize by noting that scaling the metric inversely scales K. There are different ways to define H^n concretely, one being the region of $\mathbb{R}^n$ with $x_0 > 0$ and metric $\delta_{\mu\nu}/x_0^2$, another being the set of points with $x_0 > 0$ and $\langle x, x \rangle = -1$ in M^{n+1} with a Lorentzian metric.

The sectional curvatures completely determine the Riemann tensor, but in general the Ricci tensor alone does not for manifolds of dimension greater than 3. However, the Riemann tensor is determined by the Ricci tensor together with the **Weyl curvature tensor** (AKA Weyl tensor, conformal tensor), whose definition (not reproduced here) removes all contractions of the Riemann tensor, so that it is the "trace-free part of the curvature" (i.e. all of its contractions vanish). The Weyl tensor is only defined and non-zero for dimensions $n > 3$.

The **Einstein tensor** is defined as

$$G(v, w) \equiv \mathrm{Ric}(v, w) - \frac{R}{2}g(v, w)$$

$$G_{ab} = R_{ab} - \frac{R}{2}g_{ab}.$$

If we define $G \equiv g^{ab}G_{ab}$ then we find that $R_{ab} = G_{ab} - Gg_{ab}/(n-2)$, so that the Einstein tensor vanishes iff the Ricci tensor does. Now, the Einstein tensor is symmetric, and by the spectral theorem can be diagonalized at a given point in an orthonormal basis, which also diagonalizes the Ricci tensor. In terms of the sectional curvature, we have

$$G(e_\mu, e_\mu) = - \sum_{\substack{i<j \\ i,j \neq \mu}} K(e_i, e_j).$$

Thus for a Riemannian manifold, the Einstein tensor $G(\hat{v}, \hat{v})$ applied to a unit vector twice can be viewed as $-(n-1)(n-2)/2$ times the average of the sectional curvatures of the planes orthogonal to $\hat{v}$. For an Einstein manifold, $R_{ab} = kg_{ab}$, so that $R = nk$ and thus the Einstein tensor $G_{ab} = (1 - n/2)kg_{ab}$ is also proportional to the metric tensor. Using the second Bianchi identity it can be shown (Frankel [1979] pp. 80-81) that the Einstein tensor is also "divergenceless," i.e.

$$\nabla_a G^{ab} = 0.$$

Recall that unless the metric is flat, there is no conserved quantity which can be associated with this vanishing divergence.

△ Frequent references to the divergencelessness of the Einstein tensor being related to a conserved quantity usually refer to some kind of particular context; one simple one is that in the limit of zero curvature, there is a set of conserved quantities due to the above equation.

9.3.6 *Curvature and geodesics*

Geometrically, the Ricci function $\text{Ric}(v)$ at a point $p \in M^n$ can be seen to measure the extent to which the area defined by the geodesics emanating from the $(n-1)$-surface perpendicular to v changes in the direction of v. Considering the three dimensional case in an orthonormal frame (and again dropping the hats in $\hat{e}_i$ to avoid clutter), we have

$$\text{Ric}(e_2) = \langle \check{R}(e_1, e_2)\vec{e}_2, e_1 \rangle + \langle \check{R}(e_3, e_2)\vec{e}_2, e_3 \rangle$$
$$= K(e_1, e_2) + K(e_3, e_2).$$

If we form a cube made from parallel transported vectors as we did for the first Bianchi identity, we can see that each sectional curvature term in $\text{Ric}(e_2)$ takes an edge of the cube and measures the length of the difference between the cube-aligned component of its parallel transport in the e_2 direction and the edge of the cube at a point parallel transported in the e_2 direction.

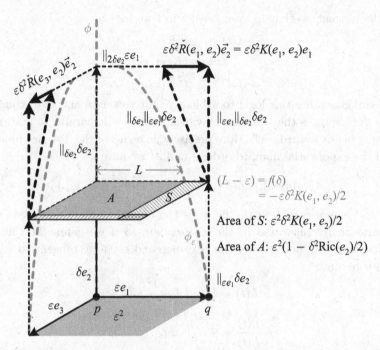

Figure 9.3.1 Each sectional curvature measures the convergence of geodesics, while their sum forms the Ricci curvature function, which measures the change in the area of the $(n-1)$-surface formed by geodesics perpendicular to its argument. In the figure we assume without loss of generality (see below) that $\check{R}(e_1, e_2)\vec{e}_2$ is parallel to e_1.

The figure above details the sectional curvature $K(e_1, e_2) = \beta^1\check{R}(e_1, e_2)\vec{e}_2$ assuming that $\check{R}(e_1, e_2)\vec{e}_2$ is parallel to e_1, so that $\langle\check{R}(e_1, e_2)\vec{e}_2, e_1\rangle = \|\check{R}(e_1, e_2)\vec{e}_2\|$. The parallel transport of e_2 along itself is depicted as parallel, so that the geodesic parametrized by arclength $\phi(t)$ is a straight line in the figure. The vector $\|_{\delta e_2}\|_{\varepsilon e_1}\,\delta e_2$ is the parallel transport of $\|_{\varepsilon e_1}\,\delta e_2$ by δ in the direction parallel to e_2, and therefore the geodesic $\phi_\varepsilon(t)$ tangent to $\|_{\varepsilon e_1}\,\delta e_2$ at q has tangent $\|_{\delta e_2}\|_{\varepsilon e_1}\,\delta e_2$ after moving a distance δ. If we consider the function $f(t)$ whose value at $t = \delta$ is the quantity $(L - \varepsilon)$ in the figure (i.e. $f(t)$ measures the offset of the geodesic from the right edge of the stack of parallel cubes), its derivative is the slope of the tangent, so that to lowest order in t we have

$$\dot{f}(t) = -\varepsilon t^2 K(e_1, e_2)/t$$
$$= -\varepsilon t K(e_1, e_2)$$
$$\Rightarrow f(t) = -\varepsilon t^2 K(e_1, e_2)/2.$$

We can generalize this logic to arbitrary unit vectors $\hat{v}$ and $\hat{w}$ to conclude that $K(\hat{v}, \hat{w})/2$ is the "fraction by which the geodesic parallel to $\hat{w}$ starting $\hat{v}$ away bends towards $\hat{w}$." More precisely, in terms of the distance function and the exponential map, to order ε and δ^2 we have

$$d\left(\exp(\delta\hat{w}), \exp(\delta \parallel_{\varepsilon\hat{v}} \hat{w})\right) = \varepsilon\left(1 - \frac{\delta^2}{2}K(\hat{v}, \hat{w})\right).$$

In the general case L in the figure is the distance between two geodesics infinitesimally separated in the $\hat{v}$ direction, so if we define $L(t)$ as this distance at any point along the parametrized geodesic tangent to $\hat{w}$, the above becomes

$$L(t) = L(0)\left(1 - \frac{t^2}{2}K(\hat{v}, \hat{w})\right)$$
$$\Rightarrow \frac{\ddot{L}(t)}{L(t)} = -K(\hat{v}, \hat{w}),$$

where the double dots indicate the second derivative with respect to t. Thus $K(\hat{v}, \hat{w})$ is "the acceleration of two parallel geodesics in the $\hat{w}$ direction with initial separation $\hat{v}$ towards each other as a fraction of the initial gap."

Now, the distance $|\varepsilon - L| = \varepsilon\delta^2 K(e_1, e_2)/2$ defines a strip S bordering the surface orthogonal to e_2 a distance δ in the e_2 direction. This strip thus has an area $\varepsilon^2\delta^2 K(e_1, e_2)/2$. If we sum this with the other strip of area $\varepsilon^2\delta^2 K(e_3, e_2)/2$, to order ε^2 and δ^2 we measure the extent to which the area A defined by the geodesics emanating from the surface perpendicular to e_2 changes in the direction of e_2. But the sum of sectional curvatures is just the Ricci function, so that in general $\mathrm{Ric}(v)/2$ is the "fraction by which the area defined by the geodesics emanating from the $(n-1)$-surface perpendicular to v changes in the direction of v." More precisely, we can follow the same logic as above, defining the "infinitesimal geodesic $(n-1)$-area" $A(t)$ along a parametrized geodesic tangent to v, so that to order ε^2 and t^2 we have

$$A(t) = \varepsilon^2\left(1 - \frac{t^2}{2}\mathrm{Ric}(v)\right)$$
$$\Rightarrow \frac{\ddot{A}(t)}{A(t)} = -\mathrm{Ric}(v).$$

Thus Ric(v) is "the acceleration of the parallel geodesics emanating from the $(n-1)$-surface perpendicular to v towards each other as a fraction of the initial surface." Note that when summing the sectional curvatures to get the Ricci function, we see that our previous assumption that $\check{R}(e_1, e_2)\vec{e}_2$ is parallel to e_1 does not affect our geometric interpretation, since the effect of any component perpendicular to e_1 is covered in the area calculation due to the other sectional curvatures. In the case of a pseudo-Riemannian manifold, "areas" and "volumes" become less geometric concepts; however, we have a clear picture in the case of a Lorentzian manifold that the Ricci function applied to a time-like vector $v \equiv \partial/\partial x^0 = \partial/\partial t$ tells us how the infinitesimal space-like volume V of free-falling particles (i.e. following geodesics) changes over time according to $\ddot{V}/V = -\text{Ric}(v) = -R_{00} = -R^\mu{}_{0\mu 0}$.

9.3.7 Jacobi fields and volumes

Now let us consider a vector field $J(t)$ along the geodesic $\phi(t)$ such that $J(0) \equiv J\big|_p = J\big|_{\phi(0)} = e_1$ and $J(\delta) \equiv J\big|_{\phi(\delta)} = (L/\varepsilon)\,\|_{\delta e_2}\, e_1$, i.e. J is the vector field "between adjacent geodesics."

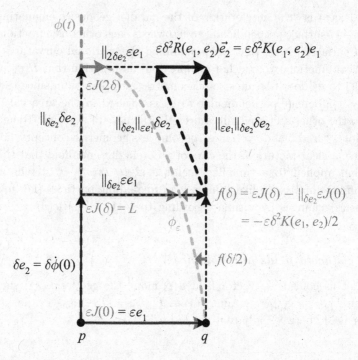

Figure 9.3.2 A Jacobi field is the vector field between adjacent geodesics, whose construction creates a relationship between the covariant derivative and the sectional curvature.

Then the function $f(t) = -\varepsilon t^2 K(e_1, e_2)/2 = -\varepsilon t^2 K(J, \dot{\phi})/2$ is the difference between J and its parallel transport in the direction tangent to ϕ, i.e. it is the value of the covariant derivative along ϕ. Since this difference is of order t^2, at $t = 0$ we have $\mathrm{D}_t^2 J = -K(J, \dot{\phi})$, or as it is more commonly written

$$\frac{\mathrm{D}^2 J}{\mathrm{d}t^2} + \check{R}(J, \dot{\phi})\vec{\phi} = 0.$$

Considered as an equation for all $J(t)$, this is called the **Jacobi equation**, with the vector field $J(t)$ that satisfies it called a **Jacobi field**. A more precise way to generalize our construction of J is to define a one-parameter family of geodesics $\phi_s(t)$, so that

$$J(t) = \left.\frac{\partial \phi_s(t)}{\partial s}\right|_{s=0}.$$

If M is complete then every Jacobi field can be expressed in this way for some family of geodesics.

If we then consider the Jacobi fields $J_v(t)$ corresponding to the geodesics $\phi_v(t)$ of tangent vectors $\|v\| = 1$ parametrized by arclength and such that to order t we have $\|J_v(1)\| = 1$, it can be shown (do Carmo [1992] pp. 114-115) that to order t^3 we have $\|J_v(t)\| = t(1 - t^2 K(J_v, \dot\phi_v)/6)$.

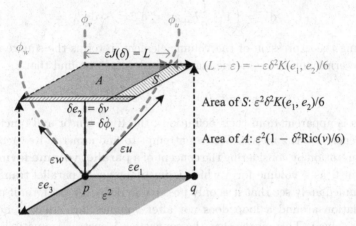

Figure 9.3.3 The infinitesimal geodesic area element derived from the Jacobi field between radial geodesics.

This means that if we apply the previous reasoning for parallel geodesics to these radial geodesics we have an "infinitesimal geodesic $(n - 1)$-area element" $A(t) = t^2(1 - t^2 \text{Ric}(v)/6)$. Integrating this over all values of v gives for small $t = \varepsilon$ the surface area of a geodesic n-ball of radius ε, which we denote $\partial B_\varepsilon(M^n)$. But this integral just averages the values of the Ricci function, which is the Ricci scalar over the dimension n, so that to order ε^2 we have

$$\frac{\partial B_\varepsilon(M^n)}{\partial B_\varepsilon(\mathbb{R}^n)} = 1 - \frac{\varepsilon^2}{6n} R,$$

and integrating over the radius we find (see A. Gray [1974]) a similar relation for the volume of a geodesic sphere compared to a Euclidean one of

$$\frac{B_\varepsilon(M^n)}{B_\varepsilon(\mathbb{R}^n)} = 1 - \frac{\varepsilon^2}{6(n + 2)} R.$$

Thus $\varepsilon^2 R/6n$ is "the fraction by which the surface area of a geodesic n-ball of radius ε is smaller than it would be under a flat metric," and $\varepsilon^2 R/6(n+2)$ is "the fraction by which the volume of a geodesic n-ball of radius ε is smaller than it would be under a flat metric."

Alternatively, we can use Riemann normal coordinates to express v in our "infinitesimal geodesic $(n-1)$-area element," whereupon following similar logic to the above we find that, at points close to the origin of our coordinates, to order $\|x\|^2$ the volume element is

$$\mathrm{d}V = \left(1 - \frac{1}{6}R_{\mu\nu}x^\mu x^\nu\right)\mathrm{d}x^1\cdots\mathrm{d}x^n,$$

or using the expression of the volume element in terms the square root of the determinant of the metric, again to order $\|x\|^2$ we find that

$$g_{\mu\nu} = \delta_{\mu\nu} - \frac{1}{3}R_{\mu\lambda\nu\sigma}x^\lambda x^\sigma.$$

As is apparent from their definitions, the Ricci tensor and function do not depend on the metric. We can attempt to find a metric-free geometric interpretation by considering the concept of a **parallel volume form**. This is defined as a volume form which is invariant under parallel translation. We immediately see that it is only possible to define such a form if parallel translation around a loop does not alter volumes, i.e. that $\check{R}$ must be $o(r, s)$-valued. This means that the connection is metric compatible, so we can define one if we wish; but if we do not, and assume zero torsion so that the Ricci tensor is symmetric, then our logic for volumes remains valid and we can still take a metric-free view of the expression for $\mathrm{d}V$ above as expressing the geodesic volume as measured by the parallel volume form. Note that unlike the Ricci tensor and function, the definitions here of the individual sectional curvatures and scalar curvature do depend upon the metric.

9.3.8 *Summary*

Below, we review the intuitive meanings of the various relations we have defined on a Riemannian manifold.

Table 9.3.4 Divergence and continuity relations and their intuitive meanings.

Relation	Meaning
$\mathrm{div}(u)\mathrm{d}V = L_u\mathrm{d}V$	$\mathrm{div}(u)$ is the fraction by which a unit volume changes when transported by the flow of u.
$\displaystyle\int_V \mathrm{div}(u)\mathrm{d}V = \int_{\partial V} i_u\mathrm{d}V$ $\displaystyle\qquad = \int_{\partial V} \langle u, \hat{n}\rangle\,\mathrm{d}S$	The change in a volume due to transport by the flow of u is equal to the net flow of u across that volume's boundary.
$\mathrm{div}(u) = 0$	u having zero divergence means the flow of u leaves volumes unchanged, or the net flow of u across the boundary of a volume is zero.
$j \equiv \rho u$, ρ is the density of Q	The current vector j is the vector whose length is the amount of Q per unit time crossing a unit area perpendicular to j
$\displaystyle\frac{\mathrm{d}q}{\mathrm{d}t} = \Sigma - \int_{\partial V} \langle j, \hat{n}\rangle\,\mathrm{d}S$	The change in q (the amount of Q within V) equals the amount generated less the amount which passes through ∂V.
$\displaystyle\frac{\partial \rho}{\partial t} = \sigma - \mathrm{div}(j)$	The change in the density of Q at a point equals the amount generated less the amount that moves away.

Table 9.3.5 Relations defined on a Riemannian manifold M^n and their intuitive meanings.

Relation	Meaning
$R \equiv g^{ab} R_{ab}$	The Ricci scalar is n times the average of the Ricci function on the set of unit tangent vectors.
$\mathrm{Ric}(e_\mu) = \sum_{i \neq \mu} g_{\mu\mu} K(e_i, e_\mu)$	The Ricci function of a unit vector is $(n-1)$ times the average of the sectional curvatures of the planes that include the vector.
$R = \sum_j g_{jj} \mathrm{Ric}(e_j)$	The Ricci scalar is n times the average of all the Ricci functions.
$R = 2 \sum_{i<j} K(e_i, e_j)$	The Ricci scalar is $n(n-1)$ times the average of all sectional curvatures.
$G(e_\mu, e_\mu) = - \sum_{\substack{i<j \\ i,j \neq \mu}} K(e_i, e_j)$	The Einstein tensor applied to a unit vector twice is $-(n-1)(n-2)/2$ times the average of the sectional curvatures of the planes orthogonal to the vector.
$d\left(\exp(\delta \hat{w}), \exp(\delta \, \|_{\varepsilon \hat{v}} \, \hat{w})\right)$ $= \varepsilon \left(1 - \dfrac{\delta^2}{2} K(\hat{v}, \hat{w})\right)$	$K(\hat{v}, \hat{w})/2$ is the fraction by which the geodesic parallel to $\hat{w}$ starting $\hat{v}$ away bends towards $\hat{w}$.
$\ddot{L}(t) = -L(t) K(\hat{v}, \hat{w})$	$K(\hat{v}, \hat{w})$ is the acceleration of two parallel geodesics in the $\hat{w}$ direction with initial separation $\hat{v}$ towards each other as a fraction of the initial gap.
$\ddot{A}(t) = -A(t) \mathrm{Ric}(v)$	$\mathrm{Ric}(v)/2$ is the fraction by which the area defined by the geodesics emanating from the $(n-1)$-surface perpendicular to v changes in the direction of v. $\mathrm{Ric}(v)$ is the acceleration of the parallel geodesics emanating from the $(n-1)$-surface perpendicular to v towards each other as a fraction of the initial surface.
$\dfrac{\partial B_\varepsilon(M^n)}{\partial B_\varepsilon(\mathbb{R}^n)} = 1 - \dfrac{\varepsilon^2}{6n} R$	$\varepsilon^2 R/6n$ is the fraction by which the surface area of a geodesic n-ball of radius ε is smaller than it would be under a flat metric.
$\dfrac{B_\varepsilon(M^n)}{B_\varepsilon(\mathbb{R}^n)} = 1 - \dfrac{\varepsilon^2}{6(n+2)} R$	$\varepsilon^2 R/6(n+2)$ is the fraction by which the volume of a geodesic n-ball of radius ε is smaller than it would be under a flat metric.

9.3.9 Related constructions and facts

A diffeomorphism that preserves angles but not lengths is called a **conformal map** (AKA conformal transformation); more precisely, a conformal map is a diffeomorphism $\Phi\colon M \to N$ such that $\forall v, w \in TM$, $\langle d\Phi_p(v), d\Phi_p(w)\rangle\big|_{\Phi(p)} = \lambda^2(p)\,\langle v, w\rangle\big|_p$, where the positive real function $\lambda^2(p)$ is called the **conformal factor**. In two dimensions, the **Riemann mapping theorem** states that any non-empty simply connected open proper subset of $\mathbb{R}^2$ can be conformally mapped to the open unit disk, a result that is not true in higher dimensions. The Weyl tensor is sometimes called the **conformal tensor** since it remains invariant under conformal mappings.

✿ Conformal maps appear in many contexts in both mathematics and physics, and can be intuitively viewed as maps which take circles to circles, although the circles may change in size. For example, the Mercator projection often used for world maps is conformal, preserving angles but distorting areas. In fact, a conformal map which preserves areas is just an isometry.

Instead of a metric, a **symplectic manifold** M is equipped with a symplectic form J (a non-degenerate 2-form) that is closed ($dJ = 0$). Symplectic manifolds arise in classical mechanics, where the points of the manifold represent the phase space of the system; this means the coordinates come in position-momentum pairs, making the dimension of the manifold even and permitting the definition of a symplectic form.

A **Hermitian manifold** is the complex version of a Riemannian manifold, a complex manifold with a (smooth) metric tensor field h, called a **Hermitian metric** (we will not go into the details of defining vectors on complex manifolds here). If we consider M as a real manifold, this metric allows us to define both a Riemannian metric $g \equiv (h + h^*)/2$, the real part of h, and a symplectic form $J \equiv i(h - h^*)/2$, the negative of the imaginary part of h. J is called the **Hermitian form** (overloading the term from Section 2.2.1, which can be applied to the Hermitian metric itself at a point). If J is closed, then M is called a **Kähler manifold** (which is also then a symplectic manifold), and J is called the **Kähler metric**. There are several ways of defining a Kähler manifold that is "orientable," which is then called a **Calabi-Yau manifold**.

Chapter 10

Fiber bundles

A manifold includes a tangent space associated with each point. A frame defines a basis for the tangent space at each point, and a connection allows us to compare vectors at different points, leading to concepts including the covariant derivative and curvature. All of these concepts can be applied to an arbitrary vector space associated with each point in place of the tangent space. This is the idea behind gauge theories. Both manifolds with connection and gauge theories can then be described using the mathematical language of fiber bundles.

10.1 Gauge theory

10.1.1 *Matter fields and gauges*

Gauge theories associate each point x on the spacetime manifold M with a (usually complex) vector space $V_x \cong \mathbb{C}^n$, called the **internal space**. A V-valued 0-form $\vec{\Phi}$ on M is called a **matter field**. A matter field lets us define analogs of the quantities from Section 9.1, as follows.

A basis for each V_x is called a **gauge**, and is the analog of a frame; choosing a gauge is sometimes called **gauge fixing**. Like the frame, a gauge is generally considered on a region $U \subseteq M$. The analog of a change of frame is then a (local) **gauge transformation** (AKA gauge transformation of the second kind), a change of basis for each V_x at each point $x \in U$. This is viewed as a representation of a **gauge group** (AKA symmetry group, structure group) G acting on V at each point $x \in U$, so that we have

$$\gamma^{-1} \colon U \to G$$

$$\rho \colon G \to GL(V)$$

$$\Rightarrow \check{\gamma}^{-1} \equiv \rho\gamma^{-1} \colon U \to GL(V),$$

and if we choose a gauge it can thus be associated with a matrix-valued
0-form or tensor field

$$(\gamma^{-1})^{\beta}{}_{\alpha} \colon U \to GL(n, \mathbb{C}),$$

so that the components of the matter field Φ^{α} transform according to

$$\Phi'^{\beta} = \gamma^{\beta}{}_{\alpha} \Phi^{\alpha}.$$

Recalling from Section 7.5.1 that all reps of a compact G are similar to a
unitary rep, for compact G we can then choose a **unitary gauge**, which is
defined to make gauge transformations unitary, so $\check{\gamma}^{-1} \colon U \to U(n)$; this is
the analog of choosing an orthonormal frame, where a change of orthonor-
mal frame then consists of a rotation at each point. A **global gauge
transformation** (AKA gauge transformation of the first kind) is a gauge
transformation that is the same at every point. If the gauge group is non-
abelian (i.e. most groups considered beyond $U(1)$), the matter field is called
a **Yang-Mills field** (AKA YM field).

$\triangle$ The term "gauge group" can refer to the abstract group G, the
matrix rep of this group within $GL(V)$, the matrix rep within $U(n)$
under a unitary gauge, or the infinite-dimensional group of maps γ^{-1}
under composition.

$\triangle$ As with vector fields, the matter field $\vec{\Phi}$ is considered to be an
intrinsic object, with only the components Φ^{α} changing under gauge
transformations.

$\triangle$ Unlike with the frame, whose global existence is determined by the
topology of M, there can be a choice as to whether a global gauge
exists or not. This is the essence of fiber bundles, as we will see in the
next section.

10.1.2 *The gauge potential and field strength*

We can then define the parallel transporter for matter fields to be a lin-
ear map $\|_{C} \colon V_p \to V_q$, where C is a curve in M from p to q. Choosing a

gauge, the parallel transporter can be viewed as a (gauge-dependent) map $\|^\beta{}_\alpha: \{C\} \to GL(n, \mathbb{C})$. This determines the (gauge-dependent) matter field connection 1-form $\Gamma^\beta{}_\alpha(v): T_xM \to gl(n, \mathbb{C})$, which can also be written when acting on a $\mathbb{C}^n$-valued 0-form as $\check{\Gamma}(v)\,\vec{\Phi}$. The values of the parallel transporter are again viewed as a rep of the gauge group G, so that the values of the connection are a rep of the Lie algebra $\mathfrak{g}$, and if G is compact we can choose a unitary gauge so that $\mathfrak{g}$ is represented by anti-hermitian matrices. We then define the **gauge potential** (AKA gauge field, vector potential) $\check{A}$ by

$$\check{\Gamma} \equiv -iq\check{A},$$

where q is called the **coupling constant** (AKA charge, interaction constant, gauge coupling parameter). Note that $A^\beta{}_\alpha$ are then hermitian matrices in a unitary gauge. The covariant derivative is then

$$\nabla_v \Phi = \mathrm{d}\vec{\Phi}(v) - iq\check{A}(v)\,\vec{\Phi},$$

which can be generalized to $\mathbb{C}^n$-valued k-forms in terms of the exterior covariant derivative as

$$D\vec{\Phi} = \mathrm{d}\vec{\Phi} - iq\check{A} \wedge \vec{\Phi}.$$

For a matter field (0-form), this is often written after being applied to e_μ as

$$D_\mu \vec{\Phi} = \partial_\mu \vec{\Phi} - iq\check{A}_\mu \vec{\Phi},$$

where μ is then a spacetime index and $\check{A}_\mu \equiv \check{A}(e_\mu)$ are $gl(n, \mathbb{C})$-valued components.

This connection defines a curvature $\check{R} \equiv \mathrm{d}\check{\Gamma} + \check{\Gamma} \wedge \check{\Gamma}$, which lets us define the **field strength** (AKA gauge field) $\check{F}$ by

$$\check{R} \equiv -iq\check{F}$$
$$\Rightarrow \check{F} = d\check{A} - iq\check{A} \wedge \check{A}.$$

10.1.3 Spinor fields

A matter field can also transform as a spinor, in which case it is called a **spinor matter field** (AKA spinor field), and is a 0-form on M which e.g. for Dirac spinors takes values in $V \otimes \mathbb{C}^4$. The gauge component then responds to gauge transformations, while the spinor component responds to changes of frame. Similarly, a matter field on M^{r+s} taking values in $V \otimes \mathbb{R}^{r+s}$ is called a **vector matter field** (AKA vector field), where the vector component responds to changes of frame. Finally, a matter field without any frame-dependent component is called a **scalar matter field** (AKA scalar field), and a matter field taking values in $\mathbb{C}$ (which can be viewed as either vectors or scalars) is called a **complex scalar matter field** (AKA complex scalar field, scalar field). A spinor matter field with gauge group $U(1)$ is called a **charged spinor field**.

$\triangle$ It is important remember that spinor and vector matter fields use the tensor product, not the direct sum, and therefore cannot be treated as two independent fields. In particular, the field value $\phi \otimes \psi \in V \otimes \mathbb{C}^4$ is identical to the value $-\phi \otimes -\psi$, which has consequences regarding the existence of global spinor fields, as we will see in Section 10.4.7.

In order to directly map changes of frame to spinor field transformations, one must use an orthonormal frame so all changes of frame are rotations. The connection associated with an orthonormal frame is therefore called a **spin connection**, and takes values in $so(3,1) \cong \mathrm{spin}(3,1)$. Thus the spin connection and gauge potential together provide the overall transformation of a spinor field under parallel translation. All of the above can be generalized to arbitrary dimension and signature.

Table 10.1.1 Constructs as applied to the various spaces associated with a point $p \in M$ in spacetime and a vector v at p.

Tangent space $T_pM = \mathbb{R}^{(r+s)}$	Spinor space $S_p = \mathbb{K}^m$	Internal space $V_p = \mathbb{C}^n$
Frame	Standard basis of $\mathbb{K}^m$ identified with an initial orthonormal frame on M	Gauge
Change of frame	$U_p \in \mathrm{Spin}(r,s)$ asssociated with change of orthonormal·frame $\breve{\gamma}_p$	Gauge transformation
Vector field $p \mapsto w \in T_pM$	Spinor field $p \mapsto \psi \in S_p$	Complex / YM field $p \mapsto \phi \in V_p$
Connection $v \mapsto \breve{\Gamma}(v) \in gl(r,s)$	Spin connection $v \mapsto \breve{\omega}(v) \in so(r,s)$, the bivectors	Gauge potential $v \mapsto \breve{A}(v) \in gl(n,\mathbb{C})$
Curvature $\breve{R} = d\breve{\Gamma} + \breve{\Gamma} \wedge \breve{\Gamma}$	Curvature $\breve{R} = d\breve{\omega} + \breve{\omega} \wedge \breve{\omega}$	Field strength $\breve{F} = d\breve{A} - iq\breve{A} \wedge \breve{A}$

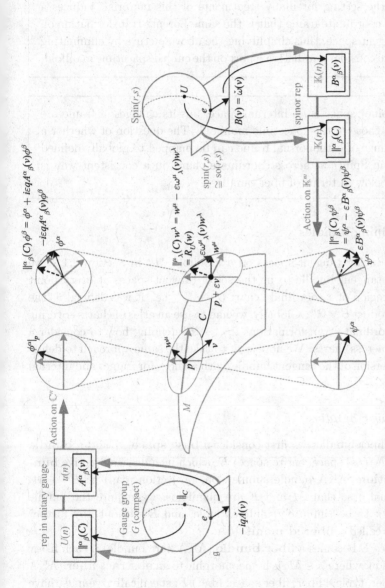

Figure 10.1.1 A matter field can be the tensor product of a complex scalar or Yang-Mills field ϕ and a spinor field ψ. YM fields use a connection and gauge (frame) which are independent of the spacetime manifold frame, while spinor fields mirror the connection and changes in frame of the spacetime manifold. YM fields are acted on by reps of the gauge group and its Lie algebra, while spinor fields are acted on by reps of the Spin group and its Lie algebra. In the figure we assume an orthonormal frame v, an infinitesimal curve C with tangent v, a spin connection, and a unitary gauge.

△ Note that a Lorentz transformation on all of flat Minkowski space, which is the setting for many treatments of this material, induces a change of coordinate frame that is the same Lorentz transformation on every tangent space, thus simplifying the above picture by eliminating the need to consider parallel transport on the curved spacetime manifold.

△ The spinor space is an internal space, but its changes of frame are driven by those of the spacetime manifold. The question of whether a global change of orthonormal frame can be mapped to globally defined elements in $\text{Spin}(r, s)$ across coordinate charts in a consistent way is resolved below in terms of fiber bundles.

10.2   Defining bundles

When introducing tangent spaces on a manifold M^n in Section 6.1.2, we defined the tangent bundle to be the set of tangent spaces at every point within the region of a coordinate chart $U \to \mathbb{R}^n$, i.e. it was defined as the cartesian product $U \times \mathbb{R}^n$. Globally, we had to use an atlas of charts covering M, with coordinate transformations $\mathbb{R}^n \to \mathbb{R}^n$ defining how to consider a vector field across charts. We now want to take the same approach to define the global version of the tangent bundle, with analogs for frames and internal spaces.

10.2.1   *Fiber bundles*

In defining fiber bundles we first consider a **base space** M and a **bundle space** (AKA total space, entire space) E, which includes a surjective **bundle projection** (AKA bundle submersion, projection map) $\pi\colon E \to M$. In the special case that M and F are manifolds, we require the bundle projections π to be (infinitely) differentiable, and E without any further structure is called a **fibered manifold**.

The space E becomes a **fiber bundle** (AKA fibre bundle) if each **fiber over** x $\pi^{-1}(x)$, where $x \in M$, is homeomorphic to an **abstract fiber** (AKA standard fiber, typical fiber, fiber space, fiber) F; specifically, we must have the analog of an atlas, a collection of open **trivializing neighborhoods** $\{U_i\}$ that cover M, each with a **local trivialization**, a homeomorphism

$$\phi_i\colon \pi^{-1}(U_i) \to U_i \times F$$
$$p \mapsto (\pi(p), f_i(p)),$$

which in a given $\pi^{-1}(x)$ allows us to ignore the first component and consider the last as a homeomorphism

$$f_i \colon \pi^{-1}(x) \to F.$$

This property of a bundle is described by calling it **locally trivial** (AKA a **local product space**), and if all of M can be made a trivializing neighborhood, then E is a **trivial bundle**, i.e. $E \cong M \times F$. The topology of a non-trivial bundle can be defined via E itself, or imputed by the local trivializations. Note that if F is discrete, then E is a covering space of M, and if M is contractible, then E is trivial.

△ Fiber bundles are denoted by various combination of components and maps in various orders, frequently (E, M, F), (E, M, π), or (E, M, π, F). Other notations include $\pi \colon E \to M$ and $F \longrightarrow E \overset{\pi}{\longrightarrow} M$.

△ The distinction between the fiber and the fiber over x is sometimes not made clear; it is important to remember that the abstract fiber F is not part of the bundle space E.

A **bundle map** (AKA bundle morphism) is a pair of maps $\Phi_E \colon E \to E'$ and $\Phi_M \colon M \to M'$ between bundles that map fibers to fibers, i.e. $\pi'(\Phi_E(p)) = \Phi_M(\pi(p))$. Note that if the bundles are over the same base space M, this reduces to a single map satisfying $\pi'(\Phi(p)) = \pi(p)$.

A **section** (AKA cross section) of a fiber bundle is a continuous map $\sigma \colon M \to E$ that satisfies $\pi(\sigma(x)) = x$. At a point $x \in M$ a **local section** always exists, being only defined in a neighborhood of x; however global sections may not exist.

△ It is important to remember that the base space M is not part of the bundle space E. In particular, since a global section may not exist, the base space cannot in general be viewed as being embedded in the bundle space, and even when it can be, such an embedding is in general arbitrary. An exception is when there is a canonical global section, for example the zero section as depicted in the Möbius strip below (and in a vector bundle in general, see Section 10.3.2).

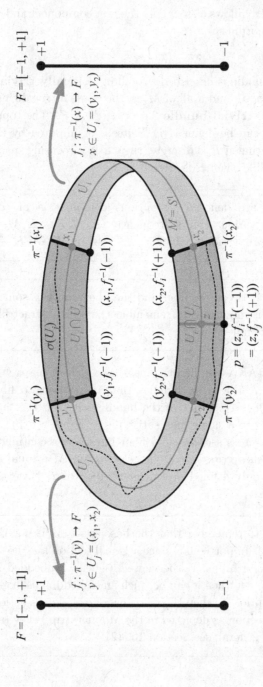

Figure 10.2.1 The **Möbius strip** (AKA Möbius band) has a base space which is a circle $M = S^1$, a fiber which is a line segment $F = [-1, +1]$, and is non-trivial, since it requires at least two trivializing neighborhoods. In the figure, the fiber over z has two different descriptions under the two local trivializations, and a local section σ (defined below) is depicted.

10.2.2   G-bundles

At a point $x \in M$ in the intersection of two trivializing neighborhoods on
a bundle (E, M, F), we have a homeomorphism $f_i f_j^{-1} \colon F \to F$. If each of
these homeomorphisms is the (left) action of an element $g_{ij}(x) \in G$, then
G is called the **structure group** of E. This action is usually required to
be faithful, so that each $g \in G$ corresponds to a distinct homeomorphism
of F. The map $g_{ij} \colon U_i \cap U_j \to G$ is called a **transition function**; the
existence of transition functions for all overlapping charts makes $\{U_i\}$ a
G-atlas and turns the bundle into a **G-bundle**. Applying the action of
g_{ij} to an arbitrary $f_j(p)$ yields

$$f_i(p) = g_{ij}\left(f_j(p)\right).$$

For example, the Möbius strip in the previous figure has a structure group
$G = \mathbb{Z}_2$, where the action of $0 \in G$ is multiplication by $+1$, and the action
of $1 \in G$ is multiplication by -1. In the top intersection $U_i \cap U_j$, $g_{ij} = 0$,
so that f_i and f_j are identical, while in the lower intersection $g_{ij} = 1$, so
that $f_i(p) = g_{ij}\left(f_j(p)\right) = 1\left(f_j(p)\right) = -f_j(p)$.

At a point in a triple intersection $U_i \cap U_j \cap U_k$, the **cocycle condition**
$g_{ij} g_{jk} = g_{ik}$ can be shown to hold, which implies $g_{ii} = e$ and $g_{ji} = g_{ij}^{-1}$.
Going the other direction, if we start with transition functions from M to G
acting on F that obey the cocycle condition, then they determine a unique
G-bundle E.

$\triangle$ It is important to remember that the left action of G is on the
abstract fiber F, which is not part of the entire space E, and whose
mappings to E are dependent upon local trivializations. A left action
on E itself based on these mappings cannot in general be consistently
defined, since for non-abelian G it will not commute with the transition
functions.

A given G-atlas may not need all the possible homeomorphisms of F
between trivializing neighborhoods, and therefore will not "use up" all
the possible values in G. If there exists trivializing neighborhoods on a
G-bundle whose transition functions take values only in a subgroup H of
G, then we say the structure group G is **reducible** to H. For example,
a trivial bundle's structure group is always reducible to the trivial group
consisting only of the identity element.

10.2.3 *Principal bundles*

A **principal bundle** (AKA principal G-bundle) (P, M, π, G) has a topological group G as both abstract fiber and structure group, where G acts on itself via left translation as a transition function across trivializing neighborhoods, i.e.

$$f_i(p) = g_{ij} f_j(p),$$

where the operation of g_{ij} is the group operation. Note that the fiber over a point $\pi^{-1}(x)$ is only homeomorphic as a space to G in a given trivializing neighborhood, and so is missing a unique identity element and is a G-torsor, not a group (see Section 7.4.1).

A principal bundle lets us introduce a consistent right action of G on $\pi^{-1}(x)$ (as opposed to the left action on the abstract fiber). This right action is defined by

$$g(p) \equiv f_i^{-1}\left(f_i(p)g\right)$$
$$\Rightarrow f_i\left(g(p)\right) = f_i(p)g$$

for $p \in \pi^{-1}(U_i)$, where in an intersection of trivializing neighborhoods $U_i \cap U_j$ we see that

$$\begin{aligned}
g(p) &= f_j^{-1}\left(f_j(p)g\right) \\
&= f_i^{-1} f_i f_j^{-1}\left(f_j(p)g\right) = f_i^{-1}\left(g_{ij} f_j(p)g\right) \\
&= f_i^{-1}\left(f_i(p)g\right) = g(p),
\end{aligned}$$

i.e. $g(p)$ is consistently defined across trivializing neighborhoods. Via this fiber-wise action, G then has a right action on the bundle P.

$\triangle$ It is important to remember that M is not part of E, and that the depiction of each fiber in the bundle $\pi^{-1}(x) \in E$ as "hovering over" the point $x \in M$ is only valid locally.

$\triangle$ Note that from its definition and basic group properties, the right action of G on $\pi^{-1}(x)$ is automatically free and transitive (making $\pi^{-1}(x)$ a "right G-torsor"). An equivalent definition of a principal bundle excludes G as a structure group but includes this free and transitive right action of G. Also note that the definition of the right action is equivalent to saying that $f_i \colon \pi^{-1}(x) \to G$ is equivariant with respect to the right action of G on $\pi^{-1}(x)$ and the right action of G on itself.

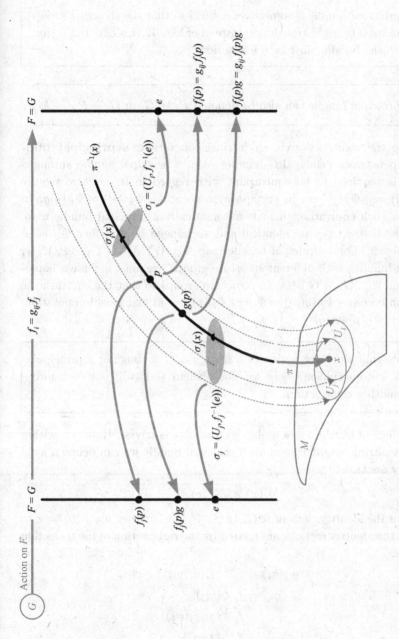

Figure 10.2.2 A principal bundle has the same group G as both abstract fiber and structure group, where G acts on itself via left translation. G also has a right action on the bundle itself, which is consistent across trivializing neighborhoods. The identity sections (defined below) are also depicted.

△ A principal bundle is sometimes defined so that the structure group acts on itself by right translation instead of left. In this case the action of G on the bundle must be a left action.

△ A principal bundle can also be denoted $P(M, G)$ or $G \hookrightarrow P \xrightarrow{\pi} M$.

Since the right action is an intrinsic operation, a **principal bundle map** between principal G-bundles (e.g. a principal bundle automorphism) is required to be equivariant with regard to it, i.e. we require $\Phi_E(g(p)) = g(\Phi_E(p))$, or in juxtaposition notation, $\Phi_E(pg) = \Phi_E(p)g$. In fact, any such equivariant map is automatically a principal bundle map, and if the base spaces are identical and unchanged by Φ_E, then Φ_E is an isomorphism. For a principal bundle map $\Phi_E \colon (P', M', G') \to (P, M, G)$ between bundles with different structure groups, we must include a homomorphism $\Phi_G \colon G' \to G$ between structure groups so that the equivariance condition becomes $\Phi_E(g(p)) = \Phi_G(g)(\Phi_E(p))$, or in juxtaposition notation, $\Phi_E(pg) = \Phi_E(p)\Phi_G(g)$.

△ Note that the right action of a fixed $g \in G$ is thus not a principal bundle automorphism, since for non-abelian G it will not commute with another right action.

A principal bundle has a global section iff it is trivial. However, within each trivializing neighborhood on a principal bundle we can define a local **identity section**

$$\sigma_i(x) \equiv f_i^{-1}(e),$$

where e is the identity element in G. In $U_i \cap U_j$, we can then use $f_i(\sigma_i) = e$ to see that the identity sections are related by the right action of the transition function:

$$
\begin{aligned}
g_{ij}(\sigma_i) &= f_i^{-1}\left(f_i(\sigma_i)g_{ij}\right) \\
&= f_i^{-1}(g_{ij}) \\
&= f_i^{-1}(g_{ij}f_j(\sigma_j)) \\
&= f_i^{-1}(f_i(\sigma_j)) \\
&= \sigma_j,
\end{aligned}
$$

or in juxtaposition notation,

$$\sigma_j = \sigma_i g_{ij}.$$

△ The different actions of G are a potential source of confusion. g_{ij} has a left action on the abstract fiber of a G-bundle, which on a principal bundle becomes left group multiplication, and also has a right action on the bundle itself that relates the elements in the identity section.

If G is a closed subgroup of a Lie group P (and thus also a Lie group by Cartan's theorem), then $(P, P/G, G)$ is a principal G-bundle with base space the (left) coset space P/G. The right action of G on the entire space P is just right translation.

10.3 Generalizing tangent spaces

In this section we use matrix notation to reduce clutter, remembering that bases are row vectors and are acted on by matrices from the right. We retain index notation when acting on vector components to avoid confusion with operations on intrinsic vectors.

10.3.1 *Associated bundles*

If two G-bundles (E, M, F) and (E', M, F'), with the same base space and structure group, also share the same trivializing neighborhoods and transition functions, then they are each called an **associated bundle** with regard to the other. It is possible to construct (up to isomorphism) a unique principal G-bundle associated to a given G-bundle; going in the other direction, given a principal G-bundle and a left action of G on a fiber F, we can construct a unique associated G-bundle with fiber F. In particular, given a principal bundle (P, M, G), the rep of G on itself by inner automorphisms defines an associated bundle $(\text{Inn}P, M, G)$, and the adjoint rep of G on $\mathfrak{g}$ defines an associated bundle $(\text{Ad}P, M, \mathfrak{g})$. If G has a linear rep on a vector space $\mathbb{K}^n$, this rep defines an associated bundle $(E, M, \mathbb{K}^n)$, which we explore next.

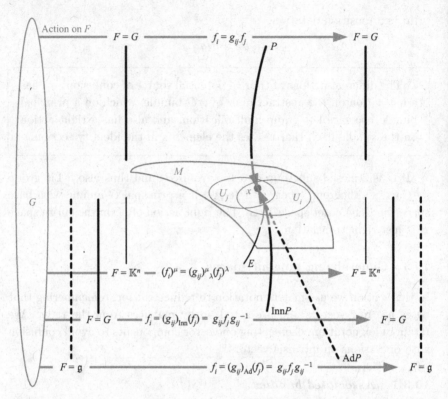

Figure 10.3.1 Given a principal bundle, we can construct an associated bundle for the action of G on a vector space $\mathbb{K}^n$ by a linear rep, on itself by inner automorphisms, and on its Lie algebra $\mathfrak{g}$ by the adjoint rep. The action of the structure group is shown in general and for the case in which G is a matrix group, with matrix multiplication denoted as juxtaposition. Although denoted identically, the f_i are those corresponding to each bundle.

> △ The G-bundle E with fiber F associated to a principal bundle P is sometimes written $E = P \times_G F \equiv (P \times F)/G$, where the quotient space collapses all points in the product space which are related by the right action of some $g \in G$ on P and the right action of g^{-1} on F.

10.3.2 Vector bundles

A **vector bundle** $(E, M, \pi, \mathbb{K}^n)$ has a vector space fiber $\mathbb{K}^n$ (assumed here to be $\mathbb{R}^n$ or $\mathbb{C}^n$) and a structure group that is linear ($G \subseteq GL(n, \mathbb{K})$) and therefore acts as a matrix across trivializing neighborhoods, i.e.

$$f_i(p) = g_{ij} f_j(p),$$

where the operation of g_{ij} is now matrix multiplication on the vector components $f_j(p) \in \mathbb{K}^n$. If we view $V_x \equiv \pi^{-1}(x)$ as an internal space on M with intrinsic vector elements v, the linear map $f_i \colon \pi^{-1}(x) \to \mathbb{K}^n$ is equivalent to choosing a basis $e_{i\mu}$ to get vector components, i.e. $f_i(v) = v_i^\mu$, where $v_i^\mu e_{i\mu} = v$ (and latin letters are labels while greek letters are the usual indices for vectors and labels for bases). The action of the structure group can then be written

$$v_i^\mu = (g_{ij})^\mu{}_\lambda v_j^\lambda,$$

which is equivalent to a change of basis

$$e_{i\mu} = (g_{ij}^{-1})^\lambda{}_\mu e_{j\lambda},$$

or as matrix multiplication on basis row vectors

$$e_j = e_i g_{ij},$$

so that the action of $g_{ij}(x)$ in $U_i \cap U_j$ is equivalent to a change of frame or gauge transformation from e_i to e_j, which is equivalent to a transformation of internal space vector components in the opposite direction.

△ The frame is not a part of the vector bundle, it is a way of viewing the local trivializations; therefore the view of $g_{ij}(x)$ as effecting a change of basis should not be confused with a group action on either $\pi^{-1}(x)$ or E. As the structure group of E, the action of G is on the fiber $\mathbb{K}^n$, which is not part of E.

If the structure group of a vector bundle is reducible to $GL(n, \mathbb{K})^e$, then it is called an **orientable bundle**; all complex vector bundles are orientable, so orientability usually refers to real vector bundles. The tangent bundle of M (formally defined in Section 10.4.5) is then orientable iff M is orientable. On a pseudo-Riemannian manifold M, the structure group of the tangent bundle is reducible to $O(r, s)$, and if M is orientable then it is reducible to $SO(r, s)$; if the structure group can be further reduced to $SO(r, s)^e$, then M and its tangent bundle are called **time and space orientable**. Note that this additional distinction is dependent only upon the metric, and two metrics on the same manifold can have different time and space orientabilities.

△ The orientability of a vector bundle as a bundle is different than its orientability as a manifold itself; therefore it is important to understand which version of orientability is being referred to. In particular, the tangent bundle of M is always orientable as a manifold, but it is orientable as a bundle only if M is.

A gauge transformation on a vector bundle is a smoothly defined linear transformation of the basis inferred by the components due to local trivializations at each point, i.e.

$$e'_{i\mu} = (\gamma_i^{-1})^\lambda{}_\mu e_{i\lambda},$$

which is equivalent to new local trivializations where

$$v'^\mu_i = (\gamma_i)^\mu{}_\lambda v^\lambda_i,$$

giving us new transition functions

$$g'_{ij} = \gamma_i g_{ij} \gamma_j^{-1},$$

where we have suppressed indices for pure matrix relationships. Thus the gauge group is the same as the structure group, and a gauge transformation γ_i^{-1} is equivalent to the transition function $g_{i'i}$ from U_i to U'_i, the same neighborhood with a different local trivialization.

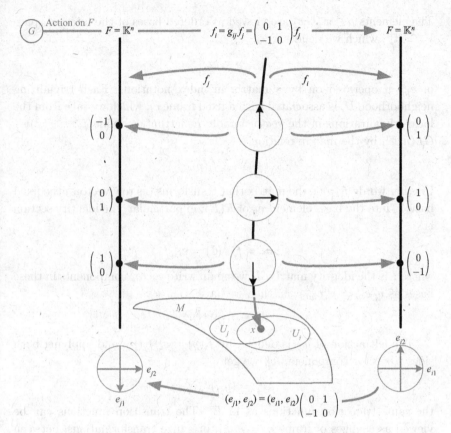

Figure 10.3.2 The elements of the fiber over x in a vector bundle can be viewed as abstract vectors in an internal space, with the local trivialization acting as a choice of basis from which the components of these vectors can be calculated. The structure group then acts as a matrix transformation between vector components, and between bases in the opposite direction. A gauge transformation is also a new choice of basis, and so can be handled similarly.

A vector bundle always has global sections (e.g. the zero vector in the fiber over each point). A vector bundle with fiber $\mathbb{R}$ is called a **line bundle**.

10.3.3 Frame bundles

Given a vector bundle $(E, M, \mathbb{K}^n)$, the **frame bundle** of E is the principal $GL(n, \mathbb{K})$-bundle associated to E, and is denoted

$$F(E) \equiv (F(E), M, \pi, GL(n, \mathbb{K})).$$

The elements $p \in \pi^{-1}(x)$ are viewed as ordered bases of the internal space $V_x \cong \mathbb{K}^n$, which we denote

$$p \equiv e_p,$$

or $e_{p\mu}$ if operated on by a matrix in index notation. Each trivializing neighborhood U_i is associated with a fixed frame e_i, which we take from the local trivializations in the vector bundle E, letting us define $f_i \colon \pi^{-1}(x) \to GL(n, \mathbb{K})$ by the matrix relation

$$e_p = e_i f_i(p).$$

In other words $f_i(p)$ is the matrix that transforms (as row vectors) the fixed basis e_i into the basis element e_p of $F(E)$; in particular, the identity section is

$$\sigma_i = f_i^{-1}(I) = e_i,$$

where I is the identity matrix. If we again write vector components in these bases as $v_i^\mu e_{i\mu} = v_p^\mu e_{p\mu} = v$, then we have

$$v_i^\mu = f_i(p)^\mu{}_\lambda v_p^\lambda.$$

The left action of g_{ij} is defined by $f_i(p) = g_{ij} f_j(p)$, and applying both sides to vector components v_p^μ we get

$$v_i^\mu = (g_{ij})^\mu{}_\lambda v_j^\lambda,$$

the same transition functions as in E. The transition functions can be viewed as changes of frame $e_j = e_i g_{ij}$, or gauge transformations, between the identity sections of $F(E)$ in $U_i \cap U_j$, i.e. this can be written as a matrix relation

$$\sigma_j = \sigma_i g_{ij},$$

which as we see next is the usual right action of the transition functions on identity sections.

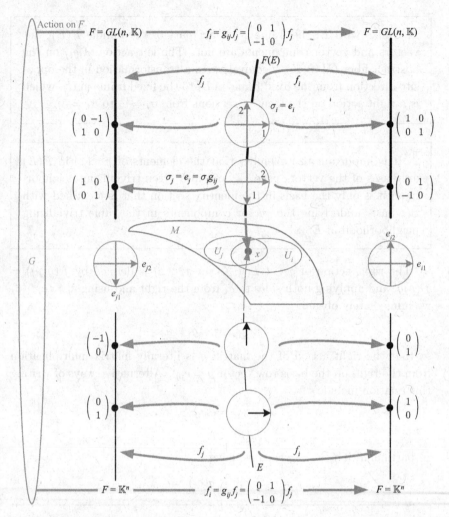

Figure 10.3.3 Given a vector bundle E, we can construct an associated frame bundle $F(E)$. The elements of the fiber over x in the frame bundle can be viewed as bases for the internal space, with the local trivialization acting as a choice of a fixed basis against which linear transformations generate these bases. These fixed bases are the same as those chosen in the corresponding local trivialization on the vector bundle, and are acted on by the same transition functions. Although denoted identically, the f_i are those corresponding to each bundle.

△ Unlike with E, the frame is in fact part of the bundle $F(E)$, but vectors and vector components are not. The left action of g_{ij} on the abstract fiber $GL(n, \mathbb{K})$ is equivalent to a transformation in the opposite direction from the fixed frame in U_i to the fixed frame in U_j, which is a right action on the identity sections from $\sigma_i = e_i$ to $\sigma_j = e_j$.

△ It is important to remember that the elements of $\pi^{-1}(x)$ in $F(E)$ are bases of the vector space V_x, and in a given trivializing neighborhood it is only the basis in the identity section that is identified with the basis underlying the vector components in the same trivializing neighborhood of E.

The right action of $g \in GL(n, \mathbb{K})$ on $\pi^{-1}(x)$ is defined by $f_i(g(p)) = f_i(p)g$, and applying both sides to e_i from the right and using $e_p = e_i f_i(p)$ we immediately obtain

$$e_{g(p)} = e_p g,$$

so that the right action of the matrix g is literally matrix multiplication from the right on the basis row vector $p = e_{p\mu}$. Alternative ways of writing this relation include

$$e_{g(p)\mu} = e_{p\mu} g^\mu{}_\lambda,$$

$$g(p) = pg.$$

In particular, if $f_i(p) = g$ then we have

$$p = e_p = e_i g = g(e_i) = e_{g(e_i)}.$$

☼ Note that since the right action on $\pi^{-1}(x)$ is by a fixed matrix, it acts as a transformation relative to each e_p, not as a transformation on the internal space V_x in which all of the bases in $\pi^{-1}(x)$ live. As a concrete example, if $g^0{}_0 = 1$ and $g^{\lambda \neq 0}{}_0 = 0$, then $e_{g(p)0} = e_{p0}$, meaning that the transformation $p \mapsto g(p)$ leaves the first component of all bases in $\pi^{-1}(x)$ unaffected. This behavior contrasts with that of a transformation on V_x itself, which as we will see in the next section is a gauge transformation.

10.3.4 *Gauge transformations on frame bundles*

Recall that a gauge transformation on a vector bundle E is an active transformation of the bases underlying the components defining a local trivialization, which is equivalent to a new set of local trivializations and transition functions (and is not a transformation on the space E itself). On the frame bundle $F(E)$, we perform the same basis change for the fixed frames associated with each trivializing neighborhood

$$e_i' = e_i \gamma_i^{-1},$$

which also defines the new identity sections, and is equivalent to new local trivializations where

$$f_i'(p) = \gamma_i f_i(p),$$

giving us new transition functions

$$g_{ij}' = \gamma_i g_{ij} \gamma_j^{-1},$$

which are the same as those in the associated vector bundle E. We will call this transformation a **neighborhood-wise gauge transformation**.

An alternative (and more common) way to view gauge transformations on $F(E)$ is to transform the actual bases in $\pi^{-1}(x)$ via a bundle automorphism

$$p' \equiv \gamma^{-1}(p),$$

and then change the fixed bases in each trivializing neighborhood to

$$e_i' = \gamma^{-1}(e_i)$$
$$\equiv e_i \gamma_i^{-1}$$

in order to leave the maps $f_i(p)$ the same (which also leaves the identity sections and transition functions the same). This immediately implies a constraint on the basis changes in $U_i \cap U_j$: since $g_{ij}' = \gamma_i g_{ij} \gamma_j^{-1}$, requiring constant g_{ij} means we must have

$$\gamma_i^{-1} = g_{ij} \gamma_j^{-1} g_{ij}^{-1}.$$

We will call this transformation an **automorphism gauge transformation**.

△ Note that this constraint means that automorphism gauge transformations are a subset of neighborhood-wise gauge transformations, which allow arbitrary changes of frame in every trivializing neighborhood. Also note that for automorphism gauge transformations, the matrices γ_i^{-1} (and therefore the new identity section elements e_i') are determined by the automorphism γ^{-1}, while neighborhood-wise gauge transformations are defined by arbitrary matrices γ_i^{-1} in each neighborhood.

✿ As with the associated vector bundle, for either type of gauge transformation the gauge group is the same as the structure group, and a gauge transformation γ_i^{-1} is equivalent to the transition function $g_{i'i}$ from U_i to U_i', the same neighborhood with a different local trivialization.

We now define the matrices γ_p^{-1} to be those which result from the transformation $\gamma^{-1}(p)$ on the rest of $\pi^{-1}(x)$, i.e.

$$e_p' \equiv e_p \gamma_p^{-1}.$$

Note that γ_p^{-1} is determined by γ_i^{-1}: since we require that $f_i' = f_i$, we have

$$
\begin{aligned}
e_i' f_i(p) &= e_p' \\
\Rightarrow e_i \gamma_i^{-1} f_i(p) &= e_p \gamma_p^{-1} \\
&= e_i f_i(p) \gamma_p^{-1} \\
\Rightarrow \gamma_p^{-1} &= f_i(p)^{-1} \gamma_i^{-1} f_i(p),
\end{aligned}
$$

or more generally, using the definition of a right action $f_i(g(p)) = f_i(p)g$ we get

$$\gamma_{g(p)}^{-1} = g^{-1} \gamma_p^{-1} g.$$

△ It is important to remember that the matrices γ_i^{-1} are dependent upon the local trivialization (since they are defined as the matrix acting on the element $e_i \in \pi^{-1}(x)$ for $x \in U_i$), but the matrices γ_p^{-1} are independent of the local trivialization, and are the action of the automorphism γ^{-1} on the basis e_p.

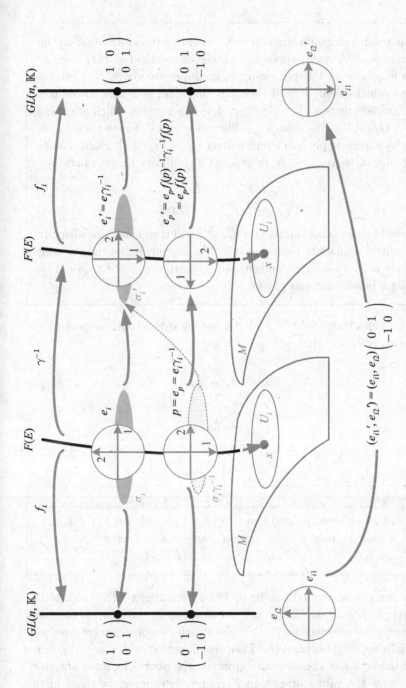

Figure 10.3.4 An automorphism gauge transformation on $F(E)$ transforms the actual elements of the fiber over x, including the identity section elements corresponding to the fixed bases in each local trivialization, thus leaving the local trivializations unchanged.

✫ This result can be understood as γ^{-1} being a transformation on the internal space V_x itself, applied to all the elements of $\pi^{-1}(x)$, each of which is a basis of V_x. For example, in the figure above, γ^{-1} rotates all bases clockwise by $\pi/2$. To see why this is so, note that the matrix in the transformation $v_i'^\mu = (\gamma_i)^\mu{}_\lambda v_i^\lambda$ has components which are those of $\gamma_i \in GL(V_x)$ in the basis $e_{i\mu}$. Therefore in a different basis $e_{p\mu} \in \pi^{-1}(x)$ we must apply a different matrix $v_p'^\mu = (\gamma_p)^\mu{}_\lambda v_p^\lambda$ which reflects the change of basis $e_{p\mu} = f_i(p)^\lambda{}_\mu e_{i\lambda}$ via a similarity transformation

$$\gamma_p = f_i(p)^{-1} \gamma_i f_i(p)$$
$$\Rightarrow \gamma_p^{-1} = f_i(p)^{-1} \gamma_i^{-1} f_i(p).$$

Viewed as a transformation on V_x, γ^{-1} will then commute with any fixed matrix applied to the bases, which as we saw is the right action; as we see next, this corresponds to the equivariance of γ^{-1} required by it being a bundle automorphism.

We now check that γ^{-1} is a bundle automorphism with respect to the right action of G, i.e. that $\gamma^{-1}(g(p)) = g\left(\gamma^{-1}(p)\right)$:

$$\gamma^{-1}(g(p)) = e_{g(p)} \gamma_{g(p)}^{-1}$$
$$= e_{g(p)} g^{-1} \gamma_p^{-1} g$$
$$= e_p \gamma_p^{-1} g$$
$$= e_{\gamma^{-1}(p)} g$$
$$= g\left(\gamma^{-1}(p)\right)$$

△ A possible source of confusion is that a local gauge transformation (different at different points) can be defined globally on $F(E)$; meanwhile, a global gauge transformation (the same matrix γ_i^{-1} at every point) can only be defined locally (unless $F(E)$ is trivial).

Consider the associated bundle to $F(E)$ with fiber $GL(\mathbb{K}^n)$, where the local trivialization of the fiber over x is defined to be the possible automorphism gauge transformations γ_i^{-1} on the identity section element over x in the trivializing neighborhood U_i. Then recalling that $\gamma_i^{-1} = g_{ij} \gamma_j^{-1} g_{ij}^{-1}$, we see that the action of the structure group on the fiber is by inner automorphism. Since the values of γ^{-1} on $F(E)$ are determined by those in the identity section, we can thus view automorphism gauge transformations as sections of the associated bundle $(\mathrm{Inn}F(E), M, GL(\mathbb{K}^n))$.

10.3.5 *Smooth bundles and jets*

Nothing we have done so far has required the spaces of a fiber bundle to be manifolds; if they are, then we require the bundle projections π to be (infinitely) differentiable and $\pi^{-1}(x)$ to be diffeomorphic to F, resulting in a **smooth bundle**. A smooth **G-bundle** then has a structure group G which is a Lie group, and whose elements correspond to diffeomorphisms of F.

If we consider a local section σ of a smooth fiber bundle (E, M, π, F) with $\sigma(x) = p$, the equivalence class of all local sections that have both $\sigma(x) = p$ and also the same tangent space $T_p\sigma$ is called the **jet** $j_p\sigma$ with **representative** σ. We can also require that further derivatives of the section match the representative, in which case the order of matching derivatives defines the **order** of the jet, which is also called a **k-jet** so that the above definition would be that of a 1-jet. x is called the **source** of the jet and p is called its **target**. With some work to transition between local sections, one can then form a **jet manifold** by considering jets with all sources and representative sections, which becomes a **jet bundle** by considering jets to be fibers over their source.

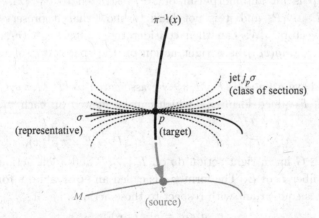

Figure 10.3.5 A jet with representative σ, source x, and target p.

10.3.6   *Vertical tangents and horizontal equivariant forms*

A smooth bundle (E, M, π) is a manifold itself, and thus has tangent vectors. A tangent vector v at $p \in E$ is called a **vertical tangent** if $\mathrm{d}\pi(v) = 0$, i.e. if it is tangent to the fiber over x where $\pi(p) = x$, so the projection down to the base space vanishes. The **vertical tangent space** V_p is then the subspace of the tangent space T_p at p consisting of vertical tangents, and viewing the vertical tangent spaces as fibers over E we can form the **vertical bundle** (VE, E, π_V), which is a subbundle of TE. We can also consider differential forms on a smooth bundle, which take arguments that are tangent vectors on E. A form is called a **horizontal form** if it vanishes whenever any of its arguments are vertical.

On a smooth principal bundle (P, M, G), we have a consistent right action $\rho \colon G \to \mathrm{Diff}(P)$, and the corresponding Lie algebra action $\mathrm{d}\rho \colon \mathfrak{g} \to \mathrm{vect}(P)$ is then a Lie algebra homomorphism. The fundamental vector fields corresponding to elements of $\mathfrak{g}$ are vertical tangent fields; in fact, at a point p, $\mathrm{d}\rho|_p$ is a vector space isomorphism from $\mathfrak{g}$ to V_p:

$$\mathrm{d}\rho|_p : \mathfrak{g} \xrightarrow{\cong} V_p$$

In addition, the right action $g \colon P \to P$ of a given element g corresponds to a right action $\mathrm{d}g \colon TP \to TP$, which maps tangent vectors on P via

$$\mathrm{d}g(v) \colon T_p P \to T_{g(p)} P.$$

This map is an automorphism of TP restricted to $\pi_P^{-1}(x)$, which we denote $T_{\pi^{-1}(x)} P$, and it is not hard to show that it preserves vertical tangent vectors. We can then consider the pullback $g^* \varphi(v_1, \ldots, v_k) = \varphi(\mathrm{d}g(v_1), \ldots, \mathrm{d}g(v_k))$ as a right action on the space $\Lambda^k P$ of k-forms on P.

If we have a bundle (E, M, π_E, F) associated to (P, M, π_P, G), we can define an F-valued form φ_P, which can be viewed on each $\pi_P^{-1}(x)$ as a mapping

$$\varphi_P \colon T_{\pi^{-1}(x)} P \otimes \cdots \otimes T_{\pi^{-1}(x)} P \to F \times \pi_P^{-1}(x),$$

where $g \in G$ has a right action $\mathrm{d}g$ on $T_{\pi^{-1}(x)} P$ and a left action g on the abstract fiber F of E. The form φ_P is called an **equivariant form** if this mapping is equivariant with respect to these actions, i.e. if

$$g^* \varphi_P = g^{-1}(\varphi_P).$$

If φ_P is also horizontal, then it is called a **horizontal equivariant form** (AKA basic form, tensorial form). If we pull back a horizontal equivariant form to the base space M using the identity sections, we get forms

$$\varphi_i \equiv \sigma_i^* \varphi_P$$

on each $U_i \subset M$. Using the identity section relation $\sigma_i = g_{ij}^{-1}(\sigma_j)$ and the pullback composition property $(g(h))^* \varphi = h^* (g^*\varphi)$, we see that the values of these forms satisfy

$$\begin{aligned} \varphi_i &= \left(g_{ij}^{-1}(\sigma_j)\right)^* \varphi_P \\ &= \sigma_j^* \left(\left(g_{ij}^{-1}\right)^* \varphi_P\right) \\ &= \sigma_j^* \left(g_{ij}\left(\varphi_P\right)\right) \\ &= g_{ij}\left(\varphi_j\right), \end{aligned}$$

where in the third line g_{ij} is acting on the value of φ_P. This means that at a point x in $U_i \cap U_j$, the values of φ_i and φ_j in the abstract fiber F correspond to a single point in $\pi_E^{-1}(x) \in E$, so that the union $\bigcup \varphi_i$ can be viewed as comprising a single E-valued form φ on M. Such a form is sometimes called a **section-valued form**, since for fixed argument vector fields its value on M is a section of E. It can be shown that the correspondence between the E-valued forms φ on M and the horizontal equivariant F-valued forms on P is one-to-one. Equivariant F-valued 0-forms on P are automatically horizontal (since one cannot pass in a vertical argument), and are thus one-to-one with sections on E.

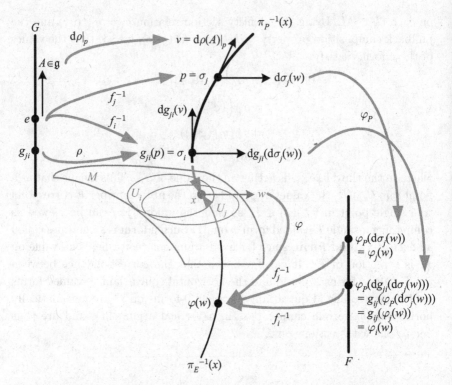

Figure 10.3.6 The differential of the right action of G on $\pi_P^{-1}(x) \in P$ creates an isomorphism to the vertical tangent space $\mathfrak{g} \cong V_p$. A horizontal equivariant form φ_P on P maps non-vertical vectors to the abstract fiber F of an associated bundle, and pulling back by the identity sections yields an E-valued form φ on M. Although denoted identically, the f_i are those corresponding to each bundle.

On the frame bundle $(P, M, \pi_P, GL(n, \mathbb{K}))$ associated with a vector bundle $(E, M, \pi_E, \mathbb{K}^n)$, a $\mathbb{K}^n$-valued form $\vec{\varphi}_P$ is then equivariant if

$$g^* \vec{\varphi}_P = \breve{g}^{-1} \vec{\varphi}_P,$$

where $\breve{g}^{-1}$ is a matrix-valued 0-form on P operating on the $\mathbb{K}^n$-valued form $\vec{\varphi}_P$. The pullback of a horizontal equivariant form on P to the base space M using the identity sections satisfies

$$\vec{\varphi}_i = \breve{g}_{ij} \vec{\varphi}_j,$$

where $\breve{g}_{ij}$ is now a matrix-valued 0-form on M. At a point x in $U_i \cap U_j$, the values of $\vec{\varphi}_i$ and $\vec{\varphi}_j$ in the abstract fiber $\mathbb{K}^n$ correspond to a single abstract vector in $V_x = \pi_E^{-1}(x) \in E$, so that the union $\bigcup \vec{\varphi}_i$ can be viewed as comprising a single V-valued form $\vec{\varphi}$ on M. Thus an equivariant $\mathbb{K}^n$-valued 0-form on P is a matter field on M.

☆ This correspondence can be viewed as follows. The right action of g on P is a transformation on bases, so that the equivalent transformation of vector components is g^{-1}. The left action of g^{-1} on the fiber is also a transformation of vector components. Thus the equivariant property can be viewed as "keeping the same value when changing basis on both bundles," so that the values of $\vec{\varphi}_P$ on $\pi_P^{-1}(x) \in P$ correspond to a single point in $\pi_E^{-1}(x) \in E$, i.e a single abstract vector over M. In other words, $\vec{\varphi} \in T_x M$ is determined by the value of $\vec{\varphi}_P$ at a single point in $\pi_P^{-1}(x) \in P$. The horizontal requirement means we do not consider forms which take non-zero values given argument vectors which project down to a zero vector on M.

Under an automorphism gauge transformation, the transformation of a horizontal equivariant form on the frame bundle P is defined by the pullback of the automorphism

$$\vec{\varphi}_P' \equiv \left(\gamma^{-1}\right)^* \vec{\varphi}_P.$$

The automorphism does not give us a right action on $T_{\pi^{-1}(x)} P$ by a fixed element, but it does give a right action when acting on the element in the identity section, so since the identity sections remain constant we have

$$\begin{aligned}
\vec{\varphi}_i' &= \sigma_i^* \left(\gamma^{-1}\right)^* \vec{\varphi}_P \\
&= \sigma_i^* \left(\gamma_i^{-1}\right)^* \vec{\varphi}_P \\
&= \sigma_i^* \check{\gamma}_i \vec{\varphi}_P \\
&= \check{\gamma}_i \vec{\varphi}_i,
\end{aligned}$$

where in the third line we used the equivariance of $\vec{\varphi}_P$. Under neighborhood-wise gauge transformations, there is no change in $\vec{\varphi}_P$ but we have new identity sections $\sigma_i'(x) = \gamma_i^{-1}(\sigma_i(x))$, so that we get

$$\begin{aligned}
\vec{\varphi}_i' &= \sigma_i'^* \vec{\varphi}_P \\
&= \left(\gamma_i^{-1}(\sigma_i)\right)^* \vec{\varphi}_P \\
&= \sigma_i^* \left(\gamma_i^{-1}\right)^* \vec{\varphi}_P \\
&= \check{\gamma}_i \vec{\varphi}_i,
\end{aligned}$$

matching the behavior for both automorphism gauge transformations and for gauge transformations as previously defined directly on M in Section 10.1.1.

Note that if a horizontal equivariant form takes values in the abstract fiber F of another bundle associated to the frame bundle, the same reasoning applies, but with $\check{\gamma}_i$ applied using the left action of G on F. In particular, recalling from Section 10.3.1 that the adjoint rep $\rho = \mathrm{Ad}$ of G on $\mathfrak{g}$ defines an associated bundle $(\mathrm{Ad}P, M, \mathfrak{g})$ to P, we can consider a $\mathfrak{g}$-valued horizontal equivariant form $\check{\Theta}_P$ on P, whose pullback by the identity section under a gauge transformation satisfies

$$\check{\Theta}'_i = \check{\gamma}_i \check{\Theta}_i \check{\gamma}_i^{-1},$$

and which similarly across trivializing neighborhoods also undergoes a gauge transformation

$$\check{\Theta}_i = \check{g}_{ij} \check{\Theta}_j \check{g}_{ij}^{-1}.$$

10.4   Generalizing connections

10.4.1   *Connections on bundles*

The fibers of a smooth bundle (E, M, π) let us define vertical tangents, but we have no structure that would allow us to canonically define a horizontal tangent. This structure is introduced via the **Ehresmann connection 1-form** (AKA bundle connection 1-form), a vector-valued 1-form on E that defines the vertical component of its argument v, which we denote v^Φ, and therefore also defines the horizontal component, which we denote $v^\ominus$:

$$\vec{\Gamma}(v) \equiv v^\Phi,$$

$$H_p \equiv \left\{ v \in T_pE \mid \vec{\Gamma}(v) = 0 \right\}$$

$$\Rightarrow v = v^\Phi + v^\ominus,$$

where $v^\Phi \in V_p$, $v^\ominus \in H_p$, and H_p is called the **horizontal tangent space**. Viewing the H_p as fibers over E then yields the **horizontal bundle** (HE, E, π_H), and a **vertical form** is defined to vanish whenever any of its arguments are horizontal. Alternatively, one can start by defining the horizontal tangent spaces as smooth sections of the jet bundle of order 1 over E, which uniquely determines a Ehresmann connection 1-form.

△ "Ehresmann connection" can refer to the horizontal tangent spaces, the horizontal bundle, the connection 1-form, or the complementary 1-form that maps to the horizontal component of its argument.

Recall that on a smooth principal bundle (P, M, π, G), the right action $\rho\colon G \to \mathrm{Diff}(P)$ has a corresponding Lie algebra action $\mathrm{d}\rho\colon \mathfrak{g} \to \mathrm{vect}(P)$ where $\mathrm{d}\rho\,|_p$ is a vector space isomorphism from $\mathfrak{g}$ to V_p. The **principal connection 1-form** (AKA principal G-connection, G-connection 1-form) is a $\mathfrak{g}$-valued vertical 1-form $\check{\Gamma}_P$ on P that defines the vertical part of its argument v at p via this isomorphism, i.e. the right action of the structure group transforms it into the Ehresmann connection 1-form:

$$\mathrm{d}\rho\left(\check{\Gamma}_P(v)\right)|_p \equiv v^\Phi$$
$$= \vec{\Gamma}(v)$$

For $g \in G$, $\mathrm{d}g(v)\colon T_pP \to T_{g(p)}P$ preserves horizontal tangent vectors as well as vertical.

$\triangle$ As with the Ehresmann connection, a "connection" on a principal bundle can refer to the principal connection 1-form, the horizontal tangent spaces, or other related quantities.

10.4.2 *Parallel transport on the frame bundle*

On a frame bundle $(P = F(E), M, \pi, GL(n, \mathbb{K}))$ with connection, we consider the horizontal tangent space to define the direction of parallel transport. More precisely, we define a **horizontal lift** of a curve C from x to y on M to be a curve C_P that projects down to C and whose tangents are horizontal:

$$\pi\left(C_P\right) = C$$
$$\dot{C}_P\,|_p \in H_p$$

There is a unique horizontal lift of C that starts at any $p \in \pi^{-1}(x)$, whose endpoint lets us define the parallel transporter on $F(E)$

$$\|_C\colon \pi^{-1}(x) \to \pi^{-1}(y).$$

It is not hard to show that the parallel transporter is a diffeomorphism between fibers, and that it commutes with the right action:

$$\|_C\left(g\left(p\right)\right) = g\left(\|_C\left(p\right)\right)$$

We can then recover the parallel transporter on M by choosing a frame (i.e. a local trivialization), using the horizontal lift that starts at the element

$\sigma_i = e_i$ in the identity section, and recalling the relation $e_p = e_i f_i(p)$:

$$\|_C (e_i \,|_x) = e_i \,|_y \, f_i \, (\|_C (e_i \,|_x))$$
$$\Rightarrow (\|_C (v))^\mu_i \,|_y = f_i \, (\|_C (e_i \,|_x))^\mu \, {}_\lambda v^\lambda_i \,|_x$$
$$\Rightarrow \|^\mu {}_\lambda (C) = f_i \, (\|_C (e_i \,|_x))^\mu {}_\lambda$$

The second line transforms vector components using the change of basis matrix in the opposite direction.

Similarly, on the frame bundle we can recover the connection 1-form on $v \in T_x M$ within a trivializing neighborhood by using the pullback of the identity section:

$$\check{\Gamma}_i(v) = \sigma_i^* \check{\Gamma}_P(v)$$
$$= \check{\Gamma}_P \, (\mathrm{d}\sigma_i(v))$$

On $F(E)$, $\sigma_i = e_i$ is the frame used to define the components of vectors in the internal space on U_i, and $\check{\Gamma}_i(v)$ then is the element of $gl(n, \mathbb{K})$ corresponding to the vertical component of v after being mapped to a tangent of the identity section. Thus since we consider the horizontal tangent space to define the direction of parallel transport, $\check{\Gamma}_i(v)$ is the infinitesimal linear transformation that takes the parallel transported frame to the frame in the direction v, the same interpretation as we found in Section 9.1.

$\triangle$ It is important to remember that $\check{\Gamma}_i$ takes values that are dependent upon the local trivialization that defines the identity section (i.e. it is frame-dependent), while the values of $\check{\Gamma}_P$ are intrinsic to the frame bundle. This reflects the fact that the connection is a choice of horizontal correspondences between frames, and so cannot have any value intrinsic to E.

The transition functions on the frame bundle can be viewed as $GL(n, \mathbb{K})$-valued 0-forms $\check{g}_{ij}$ on $U_i \cap U_j$, and it can be shown that

$$\check{\Gamma}_i(v) = \check{g}_{ij} \check{\Gamma}_j(v) \check{g}_{ij}^{-1} + \check{g}_{ij} \mathrm{d}\check{g}_{ij}^{-1}(v),$$

which is the transformation of the connection 1-form under a change of frame $\check{g}_{ij}^{-1}$ from Section 9.1.4. This is consistent with the interpretation of the action of g_{ij} as a change of frame g_{ij}^{-1} in Section 10.3.2, and it can be shown that a unique connection on $F(E)$ is determined by locally defined connection 1-forms on M and sections that are related by the same transition functions.

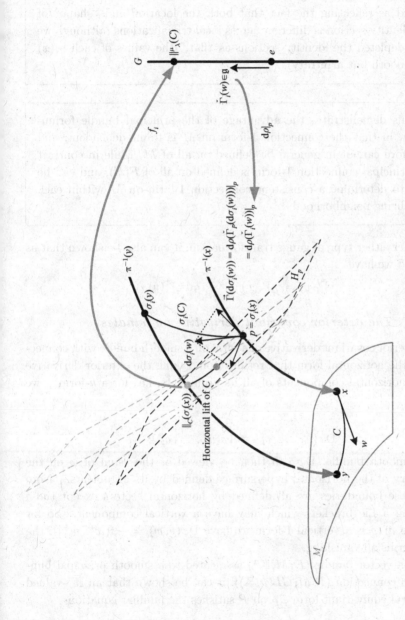

Figure 10.4.1 A principal connection 1-form on (P, M, G) defines the vertical component of its argument as a value in the Lie algebra $\mathfrak{g}$ via the isomorphism defined by the differential of the right action $d\rho$. A horizontal lift of a curve C yields the parallel transporter, and the pullback by the identity section recovers the connection 1-form on M.

☼ The inhomogeneous transformation of the connection 1-form can be viewed as reflecting the fact that both the location and "shape" of the identity section is different across local trivializations (although we have depicted the identity sections as "flat," the values of each $\sigma_i(x)$ are smooth but arbitrary).

☼ This demonstrates the advantage of the principal bundle formulation, in that the connection 1-form on M is frame-dependent, and therefore cannot in general be defined on all of M, while in contrast the principal connection 1-form is defined on all of $F(E)$, and can be used to determine a consistent connection 1-form on M within each trivializing neighborhood.

Under either type of gauge transformation, it can also be shown that as expected we have

$$\check{\Gamma}'_i(v) = \check{\gamma}_i \check{\Gamma}_i(v) \check{\gamma}_i^{-1} + \check{\gamma}_i \mathrm{d} \check{\gamma}_i^{-1}(v).$$

10.4.3 The exterior covariant derivative on bundles

The exterior covariant derivative of a form on a smooth bundle with connection is the horizontal form that results from taking the exterior derivative on the horizontal components of all its arguments, i.e. for a k-form φ we define

$$\mathrm{D}\varphi(v_0, \ldots, v_k) \equiv \mathrm{d}\varphi(v_0^\ominus, \ldots, v_k^\ominus).$$

On a smooth bundle, $\mathrm{D}\varphi$ can then be viewed as the "sum of φ on the boundary of the horizontal hypersurface defined by its arguments." Note that these boundaries are all defined by horizontal vectors except those including a Lie bracket, which may have a vertical component. So for example, if φ is a vertical 1-form we have $\mathrm{D}\varphi(v,w) = -\varphi([v^\ominus, w^\ominus])$, the other terms all vanishing.

For a vector bundle $(E, M, \mathbb{K}^n)$ associated to a smooth principal bundle with connection $(P, M, GL(n, \mathbb{K}))$, it can be shown that an $\mathbb{K}^n$-valued horizontal equivariant form $\vec{\varphi}_P$ on P satisfies the familiar equation

$$\mathrm{D}\vec{\varphi}_P = \mathrm{d}\vec{\varphi}_P + \check{\Gamma}_P \wedge \vec{\varphi}_P,$$

where as usual the derivatives are taken on the components of $\vec{\varphi}_P$, and the action of $gl(n, \mathbb{K})$-valued $\check{\Gamma}_P$ on the values of $\vec{\varphi}_P$ in is the differential of the left action of $GL(n, \mathbb{K})$. $\mathrm{D}\vec{\varphi}_P$ is then also a horizontal equivariant form. Applying the pullback by the identity section to the exterior covariant derivative, we obtain the expected

$$\mathrm{D}\vec{\varphi}_i = \mathrm{d}\vec{\varphi}_i + \check{\Gamma}_i \wedge \vec{\varphi}_i.$$

△ As with the connection 1-form, it is important to remember that the values of $\vec{\varphi}_i$ on M are components operated on by the matrix $\check{\Gamma}_i$, both of which are defined by a local trivialization.

The immediate application of the above is to a $\mathbb{K}^n$-valued form on the frame bundle. However, we can also apply it to other associated bundles to P. In particular, recalling Section 10.3.6, in the associated bundle $(\mathrm{Ad}P, M, gl(n, \mathbb{K}))$ we can apply it to a $gl(n, \mathbb{K})$-valued horizontal equivariant form $\check{\Theta}_P$ on P, where the left action of $gl(n, \mathbb{K})$ on itself is $\mathrm{d}\rho = \mathrm{ad}$, i.e. the Lie bracket. For such a form we then have

$$\mathrm{D}\check{\Theta}_P = \mathrm{d}\check{\Theta}_P + \check{\Gamma}_P[\wedge]\check{\Theta}_P,$$

where again the exterior derivative is taken on the matrix components of $\check{\Theta}_P$, and the action of $gl(n, \mathbb{K})$-valued $\check{\Gamma}_P$ on the values of $\check{\Theta}_P$ is the Lie bracket, the differential of the left action of $GL(n, \mathbb{K})$. Applying the pullback by the identity section recovers the same formula for algebra-valued forms on M, as previously seen in Section 9.2.3.

10.4.4 *Curvature on principal bundles*

On a smooth principal bundle with connection (P, M, G), the exterior covariant derivative gives us a definition for the curvature of the principal connection, the horizontal $\mathfrak{g}$-valued 2-form on P

$$\check{R}_P \equiv D\check{\Gamma}_P.$$

Recall that in Section 9.2.5 we saw that the analog of the above equation on M itself did not hold. Since $\check{\Gamma}_P$ is vertical, this can be written

$$\check{R}_P(v, w) = -\check{\Gamma}_P([v^\ominus, w^\ominus])$$

$$\Rightarrow \mathrm{d}\rho\left(\check{R}_P(v, w)\right)|_p = -[v^\ominus, w^\ominus]^\Phi,$$

so that the curvature of the principal connection is the element of $\mathfrak{g}$ corresponding to the vertical component of the Lie bracket of the horizontal components of its arguments.

On a frame bundle, we associate the horizontal tangent space with parallel transport, and the curvature is the "infinitesimal linear transformation between parallel transport in opposite directions around the boundary of the horizontal hypersurface defined by its arguments," or equivalently the "infinitesimal linear transformation associated with the vertical component of the negative Lie bracket of the horizontal components of its arguments." The curvature on M can be recovered using identity sections σ_i as with the connection:

$$\check{R}_i \equiv \sigma_i^* \check{R}_P$$

When G is a matrix group, we find analogs of equations for curvature on M using the relations from the previous section:

$$\check{R}_P = \mathrm{d}\check{\Gamma}_P + \frac{1}{2}\check{\Gamma}_P[\wedge]\check{\Gamma}_P$$

$$\check{R}_i = \mathrm{d}\check{\Gamma}_i + \frac{1}{2}\check{\Gamma}_i[\wedge]\check{\Gamma}_i$$

$$\mathrm{D}\check{R}_P = 0$$

Note that $\check{R}_P$ is a map from 2-forms on P to $\mathfrak{g}$, where G has a left action via the adjoint rep of G on $\mathfrak{g}$. One can then show that $\check{R}_P$ is equivariant with respect to this action and that of G on 2-forms, i.e. we have

$$g^*\check{R}_P = g_{\mathrm{Ad}}^{-1}\left(\check{R}_P\right).$$

Thus $\check{R}_P$ is a horizontal equivariant form, and recalling Section 10.3.6 we have the expected transformations

$$\check{R}_i = \check{g}_{ij}\check{R}_j\check{g}_{ij}^{-1},$$
$$\check{R}_i' = \check{\gamma}_i\check{R}_i\check{\gamma}_i^{-1}.$$

If a flat connection (zero curvature) can be defined on a principal bundle, then the structure group is discrete. If in addition the base space is simply connected, then the bundle is trivial.

10.4.5 *The tangent bundle and solder form*

Returning to our motivating example, the **tangent bundle** on a manifold M^n, denoted TM, is a smooth vector bundle $(E, M^n, \mathbb{R}^n)$ with a (possibly reducible) structure group $GL(n, \mathbb{R})$ that acts as an inverse change of local frame across trivializing neighborhoods. These trivializing neighborhoods can be obtained from an atlas on M, with fiber

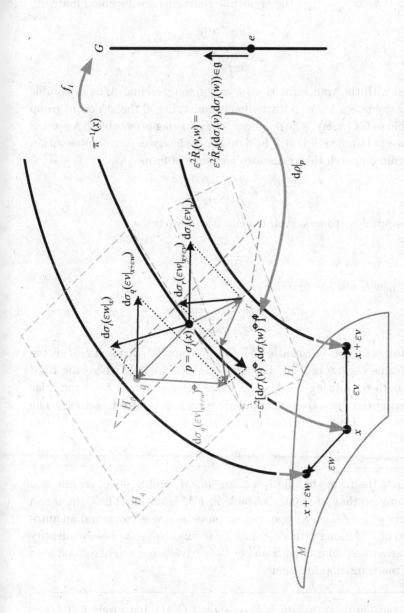

Figure 10.4.2 The curvature of the principal connection is the element of $\mathfrak{g}$ corresponding to the vertical component of the negative Lie bracket of the horizontal components of its arguments. The sections used at q and r are arbitrary, since they don't affect the vertical component of the loop remainder. If the arguments are pulled back using the identity section, we recover the curvature on the base space M.

homeomorphisms $f_i\colon T_xM \to \mathbb{R}^n$ defined by components in the coordinate frame $e_{i\mu} = \partial/\partial x_i^\mu$, so that the transition functions are Jacobian matrices

$$v_i^\mu = (g_{ij})^\mu{}_\lambda v_j^\lambda$$

$$= \frac{\partial x_i^\mu}{\partial x_j^\lambda} v_j^\lambda$$

associated with the transformation of vector components. M is orientable iff these Jacobians all have positive determinant, i.e. iff the structure group is reducible to $GL(n, \mathbb{R})^e$ (the definition of TM being orientable). A section of the tangent bundle is a vector field on M. A change of coordinates within each coordinate patch then generates a change of frame

$$\frac{\partial}{\partial x_i'^\mu} = \frac{\partial x_i^\mu}{\partial x_i'^\lambda} \frac{\partial}{\partial x_i^\lambda},$$

which is equivalent to new local trivializations where

$$v_i'^\mu = \frac{\partial x_i'^\mu}{\partial x_i^\lambda} v_i^\lambda,$$

giving us new transition functions

$$\frac{\partial x_i'^\mu}{\partial x_j'^\lambda} = \frac{\partial x_i'^\mu}{\partial x_i^\sigma} \frac{\partial x_i^\sigma}{\partial x_j^\nu} \frac{\partial x_j^\nu}{\partial x_j'^\lambda}.$$

The **tangent frame bundle** (AKA frame bundle), denoted FM, is the smooth frame bundle of TM, i.e. $(FM, M^n, GL(n, \mathbb{R}))$, where the fixed bases in each trivializing neighborhood are again obtained from the atlas on M, giving the same transition functions as in the tangent bundle. The bases in $\pi^{-1}(x)$ are thus defined by

$$e_{p\mu} = f_i(p)^\lambda{}_\mu \frac{\partial}{\partial x_i^\lambda}.$$

A section of the frame bundle is a frame on M, and a global section is a global frame, so that M is parallelizable iff FM is trivial. The right action of a matrix $g^\mu{}_\lambda \in GL(n, \mathbb{R})$ operates on bases as row vectors, and an automorphism of FM along with a redefinition of fixed bases to preserve identity sections generates changes of frame in each trivializing neighborhood that preserve the transition functions.

$\triangle$ The tangent frame bundle is also denoted $F(M)$, but rarely $F(TM)$, which is what would be consistent with general frame bundle notation.

The tangent frame bundle is special in that we can relate its tangent vectors to the elements of the bundle as bases. Specifically, we define the **solder form** (AKA soldering form, tautological 1-form, fundamental 1-form), as a $\mathbb{R}^n$-valued 1-form $\vec{\theta}_P$ on $P = FM^n$ which at a point $p = e_p$ projects its argument $v \in T_pFM$ down to M and then takes the resulting vector's components in the basis e_p, i.e.

$$\vec{\theta}_P(v) \equiv \mathrm{d}\pi(v)_p^\mu.$$

The projection makes the solder form horizontal, and it is also not hard to show it is equivariant, since both actions essentially effect a change of basis:

$$g^*\vec{\theta}_P(v) = \breve{g}^{-1}\vec{\theta}_P(v).$$

The pullback by the identity section

$$\vec{\theta}_i \equiv \sigma_i^*\vec{\theta}_P$$

simply returns the components of the argument in the local basis, and thus is identical to the dual frame $\vec{\beta}$ viewed as a frame-dependent $\mathbb{R}^n$-valued 1-form from Section 9.2.4. Thus recalling Section 10.3.6, the values of $\vec{\theta}_P$ in the fiber over x correpond to a single point in the associated bundle TM, so that the union of the pullbacks $\vec{\theta}_i$ can be viewed as a single TM-valued 1-form on M

$$\vec{\theta} \colon TM \to TM$$

which identifies, or "solders," the tangent vectors on M to elements in the bundle TM associated to FM (explaining the alternative name "tautological 1-form").

$\triangle$ The TM-valued 1-form $\vec{\theta}$ is also sometimes called the solder form, and can be generalized to bundles E with more general fibers as $\theta_E(v) \colon TM \to E$ or $\theta_{\sigma_0}(v) \colon TM \to V_{\sigma_0}E$, where in the second case σ_0 is a distinguished section (e.g. the zero section in a vector bundle). This is called a **soldering** of E to M; for example a Riemannian metric provides a soldering of the cotangent bundle to M. In classical dynamics, if M is a configuration space then the solder form to the cotangent bundle is called the Liouville 1-form, Poincaré 1-form, canonical 1-form, or symplectic potential.

> △ The solder form can also be used to identify the tangent space with a subspace of a vector bundle over M with higher dimension than M.

10.4.6   *Torsion on the tangent frame bundle*

The covariant derivative of the solder form defines the torsion on P

$$\vec{T}_P \equiv D\vec{\theta}_P$$
$$= d\vec{\theta}_P + \check{\Gamma}_P \wedge \vec{\theta}_P.$$

$\vec{T}_P$ is a horizontal equivariant form since $\vec{\theta}_P$ is. Examining the first few components, we have:

$$\vec{T}_P(v, w) = d\vec{\theta}_P(v^\ominus, w^\ominus)$$
$$\Rightarrow \varepsilon^2 \vec{T}_P(v, w) = \vec{\theta}_P(\varepsilon w^\ominus|_{p+\varepsilon v^\ominus}) - \vec{\theta}_P(\varepsilon w^\ominus|_p) - \dots$$
$$= d\pi(\varepsilon w^\ominus|_{p+\varepsilon v^\ominus})^\mu_{p+\varepsilon v^\ominus} - d\pi(\varepsilon w^\ominus|_p)^\mu_p - \dots$$

The first term projects the horizontal component of w at $p + \varepsilon v^\ominus$ down to M, which is the same as projecting w itself down to M since the projection of the vertical part vanishes. Then we take its components in the basis at $p + \varepsilon v^\ominus$, which is the parallel transport of the basis at p in the direction v. These are the same components as that of the projection of w at $p + \varepsilon v^\ominus$ parallel transported back to p in the basis at p. Thus the torsion on P is the "sum of the boundary vectors of the surface defined by the projection of its arguments down to M after being parallel transported back to p."

This analysis makes it clear that the pullback of the torsion on P by the identity section

$$\vec{T}_i \equiv \sigma_i^* \vec{T}_P,$$

which by our previous pullback results recovers the torsion on M, just bounces the argument vectors to the identity section and back, thus yielding the same interpretation for torsion that we arrived at in Section 9.2.4.

It can also be shown that the analog of the first Bianchi identity M holds on P, with the original being recovered upon pulling back by the identity section:

$$D\vec{T}_P = \check{R}_P \wedge \vec{\theta}_P$$
$$D\vec{T}_i = \check{R}_i \wedge \vec{\theta}_i$$

10.4.7 *Spinor bundles*

A **spin structure** on an orientable Riemannian manifold M is a principal bundle map

$$\Phi_P \colon (P, M^n, \mathrm{Spin}(n)) \to (F_{SO}, M^n, SO(n))$$

from the **spin frame bundle** (AKA bundle of spin frames) P to the orthonormal frame bundle F_{SO} with respect to the double covering map $\Phi_G \colon \mathrm{Spin}(n) \to SO(n)$. The equivariance condition on the bundle map is then $\Phi_P(U(p)) = \Phi_G(U)(\Phi_P(p))$, so that the right action of a spinor transformation $U \in \mathrm{Spin}(n)$ on a spin basis corresponds to the right action of a rotation $\Phi_G(U)$ on the corresponding orthonormal basis $\Phi_P(p)$. On a time and space orientable pseudo-Riemannian manifold, a spin structure is a principal bundle map with respect to the double covering map $\Phi_G \colon \mathrm{Spin}(r, s)^e \to SO(r, s)^e$ (except in the case $r = s = 1$, which is not a double cover).

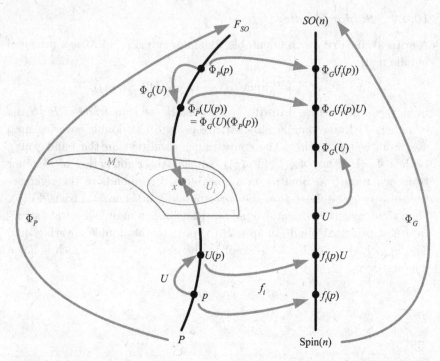

Figure 10.4.3 A spin structure is a principal bundle map that gives a global 2-1 mapping from the fibers of the spin frame bundle to the fibers of the orthonormal frame bundle. The existence of a spin structure means that a change of frame can be smoothly and consistently mapped to changes of spin frame, permitting the existence of spinor fields.

If a spin structure exists for M, then M is called a **spin manifold** (one also says M is spin; sometimes a spin manifold is defined to include a specific spin structure). Any manifold that can be defined with no more than two coordinate charts is then spin, and therefore any parallelizable manifold and any n-sphere is spin. As we will see in Section 10.5.2, the existence of spin structures can be related to characteristic classes. It also can be shown that any non-compact spacetime manifold with signature $(3, 1)$ is spin iff it is parallelizable. Finally, a vector bundle $(E, M^n, \mathbb{C}^m)$ associated to the spin frame bundle $(P, M, \mathrm{Spin}(r, s)^e)$ under a rep of $\mathrm{Spin}(r, s)^e$ on $\mathbb{C}^m$ is called a **spinor bundle**, and a section of this bundle is a spinor field on M.

For a charged spinor field taking values in $U(1) \otimes \mathbb{C}^m$, where $\mathbb{C}^m$ is acted on by a rep of $\mathrm{Spin}(r, s)^e$, the action of $(e^{i\theta}, U) \in U(1) \times \mathrm{Spin}(r, s)^e$ and

$(-e^{i\theta}, -U)$ are identical, so that the structure group is reducible to

$$\mathrm{Spin}^c(r, s)^e \equiv U(1) \times_{\mathbb{Z}_2} \mathrm{Spin}(r, s)^e$$
$$\equiv (U(1) \times \mathrm{Spin}(r, s)^e)/\mathbb{Z}_2,$$

where the quotient space collapses all points in the product space which are related by changing the sign of both components. The superscript refers to the circle $U(1)$. A **spin^c structure** on an orientable pseudo-Riemannian manifold M is then a principal bundle map

$$\Phi_P \colon (P, M^n, \mathrm{Spin}^c(r, s)^e) \to (F_{SO}, M^n, SO(r, s)^e)$$

with respect to the double covering map $\Phi_G \colon \mathrm{Spin}^c(r, s)^e \to SO(r, s)^e$ where the $U(1)$ factor is ignored. For spinor matter fields that take values in $V \otimes \mathbb{C}^m$ for some internal space V with structure (gauge) group G with $\mathbb{Z}_2$ in its center (e.g. a matrix group where the negative of every element remains in the group), we can analogously define a **spin^G structure**. It can be shown (see Avis and Isham [1980]) that spin^G structures exist on any four dimensional M if such a G is a compact simple simply connected Lie group, e.g. $SU(2i)$; therefore the spacetime manifold has no constraints due to spin structure in the standard model, or in any extension that includes $SU(2)$ gauged spinors.

10.5 Characterizing bundles

10.5.1 *Universal bundles*

Given a fiber bundle (E, M, π, F) and a continuous map to the base space $f \colon N \to M$, the **pullback bundle** (AKA induced bundle, pullback of E by f) is defined as

$$f^*(E) \equiv \{(n, p) \in N \times E \mid f(n) = \pi(p)\},$$

and is a fiber bundle $(f^*(E), N, \pi_f, F)$ with the same fiber but base space N. Projection of $q = (n, p) \in f^*(E)$ onto n is just the bundle projection $\pi_f \colon f^*(E) \to N$, while projection onto p defines a bundle map $\Phi \colon f^*(E) \to E$ such that $\pi(\Phi(q)) = f(\pi_f(q)) = x \in B$.

For any topological group G, there exists a **universal principal bundle** (AKA universal bundle) (EG, BG, G) such that every principal G-bundle (P, M, G) (with M at least a CW-complex) is the pullback of EG by some $f \colon M \to BG$. The base space BG is called the **classifying space** for G. The pullbacks of a principal bundle by two homotopic maps are isomorphic, and thus for a given M the homotopy classes of the maps f are one-to-one with the isomorphism classes of principal G-bundles over M.

Every vector bundle $(E, M, \mathbb{K}^n)$ is therefore the pullback of the **universal vector bundle** $E_n(\mathbb{K}^\infty)$ (AKA tautological bundle, universal bundle), the vector bundle associated to the universal principal bundle for its structure group. It can be shown that any vector bundle admits an inner product, so we need only consider the structure groups $O(n)$ and $U(n)$, whose classifying spaces are each a **Grassmann manifold** (AKA Grassmannian) $G_n(\mathbb{K}^\infty)$. This is a limit of the finite-dimensional Grassmann manifold $G_n(\mathbb{K}^k)$, which is all n-planes in $\mathbb{K}^k$ through the origin. Each point $x \in G_n(\mathbb{K}^k)$ thus corresponds to a copy of $\mathbb{K}^n$, as does the fiber over x in the universal vector bundle, explaining the alternate name "tautological bundle." The total space of the associated universal principal bundle is the **Stiefel manifold** $V_n(\mathbb{K}^\infty)$, a limit of the finite-dimensional $V_n(\mathbb{K}^k)$, defined as all ordered orthonormal n-tuples in $\mathbb{K}^k$; the bundle projection simply sends each n-tuple to the n-plane containing it.

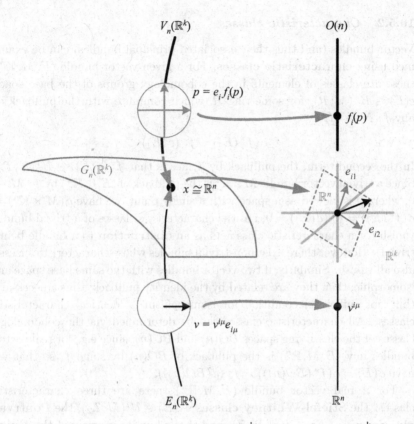

Figure 10.5.1 The Grassmann manifold $G_n(\mathbb{R}^k)$ is all n-planes in $\mathbb{R}^k$ through the origin, and is the base space of the Stiefel manifold $V_n(\mathbb{R}^k)$, defined as all ordered orthonormal n-tuples in $\mathbb{R}^k$ where the fiber is $O(n)$ and the bundle projection simply sends each n-tuple to the n-plane containing it. The tautological bundle is the associated vector bundle $E_n(\mathbb{R}^k)$ with fiber $\mathbb{R}^n$, and the universal principal bundle for $O(n)$ is the limit $V_n(\mathbb{K}^\infty)$.

$\triangle$ Grassmann manifolds can also be denoted $Gr(n, \mathbb{K}^k)$, $Gr(n, k)$, $G_{n,k}$ or $g_{n,k}$ and the order of the parameters are sometimes reversed. Stiefel manifolds have similar alternative notations.

10.5.2   *Characteristic classes*

Vector bundles (and thus their associated principal bundles) can be examined using **characteristic classes**. For a given vector bundle $(E, M, \mathbb{K}^n)$ these are classes of elements in the cohomology groups of the base space $c(E) \in H^*(M; R)$, for some ring R, which commute with the pullback of any $f: N \to M$:

$$c(f^*(E)) = f^*(c(E))$$

In the second term, the pullback by f means that $f^*(c(E)) \in H^*(N; R)$. Since a trivial vector bundle $M \times \mathbb{K}^n$ is the pullback of $(E, 0, \mathbb{K}^n)$ by $f: M \to 0$, where 0 is viewed as a space with a single point, we have $c(M \times \mathbb{K}^n) = c(f^*(E)) = f^*(c(E)) = 0$, i.e. the characteristic classes of a trivial bundle vanish (or a characteristic class acts as an **obstruction** to a bundle being trivial). However there exist non-trivial bundles whose characteristic classes also all vanish. Similarly, if two vector bundles with the same base space are isomorphic, then they are related by the identity pullback; thus a necessary (but not sufficient) condition for isomorphism is identical characteristic classes. All characteristic classes can be determined via the cohomology classes of the classifying spaces $BO(n)$ and $BU(n)$, since e.g. for real vector bundles any $(E, M, \mathbb{R}^n)$ is the pullback of $BO(n)$ by some f, so that we have $c(E) = c(f^*(BO(n))) = f^*(c(BO(n)))$.

For a real vector bundle $(E, M, \mathbb{R}^n)$ there are three characteristic classes: the **Stiefel-Whitney classes** $w_i(E) \in H^i(M; \mathbb{Z}_2)$, the **Pontryagin classes** $p_i(E) \in H^{4i}(M; \mathbb{Z})$, and if the bundle is oriented the **Euler class** $e(E) \in H^n(M; \mathbb{Z})$. For complex vector bundles, there are the **Chern classes** $c_i(E) \in H^{2i}(M; \mathbb{Z})$. The characteristic class of a manifold M is defined to be that of its tangent bundle, e.g.

$$w_i(M) \equiv w_i(TM).$$

If M is a compact orientable four-dimensional manifold, then it is parallelizable iff $w_2(M) = p_1(M) = e(M) = 0$.

A non-zero Stiefel-Whitney class $w_i(E)$ acts as an obstruction to the existence of $(n - i + 1)$ everywhere linearly independent sections of E. Therefore, if such section do exist, then $w_j(E)$ vanishes for $j \geq i$; in particular, a non-zero $w_n(E)$ means there are no non-vanishing global sections. It can be shown that $w_1(E) = 0$ iff E is orientable, so that M is orientable iff $w_1(M) = 0$.

Spin structures exist on an oriented M iff $w_2(M) = 0$; if spin structures do exist, then their equivalence classes have a one-to-one correspondence with the elements of $H^1(M, \mathbb{Z}_2)$. Inequivalent spin structures have either inequivalent spin frame bundles or inequivalent bundle maps; in four dimensions, there is only one spin frame bundle up to isomorphism, so that different spin structures correspond to different bundle maps (i.e. different spin connections).

Spin^c structures exist on an oriented M if spin structures exist, but also in some cases where they do not; for example if M is simply connected and compact. If spin^c structures do exist, then their equivalency classes have a one-to-one correspondence with the elements of $H^2(M, \mathbb{Z})$, and in four dimensions, unlike the case for spin structures, inequivalent spin^c structures can have inequivalent spin frame bundles.

10.5.3 *Related constructions and facts*

The direct product of two vector bundles $(E, M, \mathbb{K}^m)$ and $(E', M', \mathbb{K}^n)$ is another vector bundle

$$(E \times E', M \times M', \mathbb{K}^{m+n}).$$

If we form the direct product of two vector bundles with the same base space, we can then restrict the base space to the diagonal via the pullback by $f \colon M \times M \to M$ defined by $(x, x) \mapsto x$. The resulting vector bundle is called the **Whitney sum** (AKA direct sum bundle), and is denoted

$$(E \oplus E', M, \mathbb{K}^{m+n}).$$

The **total Whitney class** of a real vector bundle $(E, M, \mathbb{R}^n)$ is defined as

$$w(E) \equiv 1 + w_1(E) + w_2(E) + \cdots + w_n(E).$$

The series is finite since $w_i(E)$ vanishes for $i > n$, and is thus an element of $H^*(M, \mathbb{Z}_2)$. The total Whitney class is multiplicative over the Whitney sum, i.e.

$$w(E \oplus E') = w(E)w(E').$$

The **total Chern class** is defined similarly, and has the same multiplicative property.

The **flag manifold** $F_n(\mathbb{K}^\infty)$ is a limit of the finite-dimensional flag manifold $F_n(\mathbb{K}^k)$, which is all ordered n-tuples of orthogonal lines in $\mathbb{K}^k$ through the origin. The name is due to the fact that an ordered n-tuple of orthogonal lines in $\mathbb{K}^k$ is equivalent to an **n-flag**, a sequence of subspaces $V_1 \subset \cdots \subset V_n$ in $\mathbb{K}^k$ where each V_i has dimension i.

Appendix A

Categories and functors

Here we provide a very brief overview of the most basic ideas in category theory, a general way to view mathematical objects and their interrelationships.

A.1 Generalizing sets and mappings

Recall the types of set mappings from Section 1.2. It can be useful to generalize both sets and mappings, which takes us from set theory to **category theory**. Category theory eliminates any dependence upon elements, referring only to classes of generic **objects**. The **class** (AKA collection) of objects ob(C) of a category C sometimes may be defined as sets with a certain structure, but in category theory they are left completely abstract, with the following definitions built upon them:

- **Morphisms**: a set mor(X, Y) (also denoted hom(X, Y) or $C(X, Y)$) of morphisms between X and Y is defined for every $X, Y \in$ ob(C); every mor(X, X) includes an identity 1_X
- **Composition**: an operator $\circ$ is defined between morphisms that is distributive and respects the identity, i.e. for morphisms $m : X \to Y$ and $n : Y \to Z$ we have $n \circ m : X \to Z$ with $(n \circ m) \circ l = n \circ (m \circ l)$ and $m \circ 1_X = m = 1_Y \circ m$

A **category** C then consists of a class of objects ob(C), a collection of sets of morphisms mor(X, Y) between these objects, and a morphism composition operator. It is helpful in understanding these definitions to consider their application to sets and mappings. In this case, a class of objects would consist of sets along with a structure; morphisms would be mappings between these objects; and the category would consist of the class and the mappings.

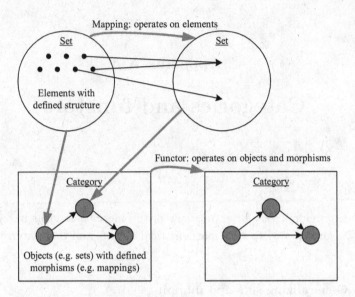

Figure A.1.1 Categories contrasted with sets as a particular example of objects.

A.2   Mapping mappings

As described so far, category theory has essentially provided an abstract generalization of sets and mappings that ignores elements and thus can be used for diverse constructs. The real power of categories comes in the next step, in which we "move up a level" and define mappings between entire categories.

- **Covariant functor**: a mapping F between categories C and D such that $\forall X \in C$, $F(X) \in D$, and $\forall m \in \mathrm{mor}(X, Y)$, $F(m) \in \mathrm{mor}\,(F(X), F(Y))$, with $F(1_X) = 1_{F(X)}$ and $F(n \circ m) = F(n) \circ F(m)$
- **Contravariant functor**: A "reversing" functor with properties as above, except that $F(m) \in \mathrm{Mor}\,(F(Y), F(X))$ and $F(n \circ m) = F(m) \circ F(n)$, i.e. a contravariant functor "flips the direction of morphisms"

To get a concrete sense of how functors work, we give an example that assumes familiarity with groups.

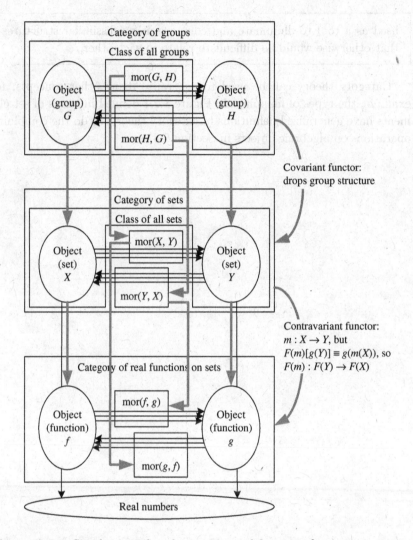

Figure A.2.1 Simple examples of categories and functors: from groups to sets to real functions on sets.

△ It should be emphasized that although we are using sets and mappings as examples of objects and morphisms, category theory can be

used as a tool to illuminate and compare many dissimilar structures
that otherwise would be difficult to relate to each other.

Category theory can be used to generalize many other concepts; for
example, the types of mappings in Figure 1.2.1 defined in terms of set elements have generalized definitions in category theory, as do the combining
operations on algebraic objects in Section 2.3.

Bibliography

Alfred A. Gray. The volume of a small geodesic ball of a riemannian manifold. *The Michigan Mathematical Journal*, 20(4):329–344, 1974. http://projecteuclid.org/euclid.mmj/1029001150.

S. J. Avis and C. J. Isham. Generalized spin structures on four dimensional space-times. *Commun. Math. Phys.*, 72:103–118, 1980. https://projecteuclid.org/euclid.cmp/1103907653.

J. Baez and J. P. Muniain. *Gauge Fields, Knots, and Gravity*. World Scientific, 1994.

K. Thorne C. Misner and J. Wheeler. *Gravitation*. W H Freeman and Company, 1973.

M. do Carmo. *Riemannian Geometry*. Birkhäuser Boston, 1992.

C. Doran and A. Lasenby. *Geometric Algebra for Physicists*. Cambridge University Press, 2003.

T. Frankel. *Gravitational Curvature*. W H Freeman and Company, 1979.

T. Frankel. *The Geometry of Physics*. Cambridge University Press, 1997.

J. Fuchs and C. Schweigert. *Symmetries, Lie Algebras and Representations: A Graduate Course for Physicists*. Cambridge University Press, 1997.

R. Geroch. *Mathematical Physics*. University of Chicago Press, 1985.

M. Göckeler and T. Schücker. *Differential Geometry, Gauge Theories, and Gravity*. Cambridge University Press, 1987.

Jr. H. B. Lawson and M. Michelsohn. *Spin Geometry*. Princeton University Press, 1989.

A. Hatcher. *Algebraic Topology*. Cambridge University Press, 2002.

A. Hatcher. *Vector Bundles and K-Theory*. Self Published, 2003. https://www.math.cornell.edu/ hatcher/VBKT/VBpage.html.

S. Kobayashi and K. Nomizu. *Foundations of Differential Geometry*. John Wiley & Sons, 1963.

P. Petersen. *Riemannian Geometry*. Springer, 2006.

R. Wald. *General Relativity*. University of Chicago Press, 1984.

F. W. Warner. *Foundations of Differentiable Manifolds and Lie Groups*. Springer-Verlag, 1983.

Index

Printed in the United States
By Bookmasters